Moduli Spaces and Arithmetic Dynamics

Volume 30

CRM MONOGRAPH SERIES

Centre de Recherches Mathématiques
Montréal

Moduli Spaces and Arithmetic Dynamics

Joseph H. Silverman

The Centre de Recherches Mathématiques (CRM) of the Université de Montréal was created in 1968 to promote research in pure and applied mathematics and related disciplines. Among its activities are special theme years, summer schools, workshops, postdoctoral programs, and publishing. The CRM is supported by the Université de Montréal, the Province of Québec (FQRNT), and the Natural Sciences and Engineering Research Council of Canada. It is affiliated with the Institut des Sciences Mathématiques (ISM) of Montréal. The CRM may be reached on the Web at www.crm.math.ca.

American Mathematical Society
Providence, Rhode Island USA

The author's research was supported by NSF DMS-0650017 and DMS-0854755.

2000 *Mathematics Subject Classification.* Primary 37P45; Secondary 37A45, 37F45, 37P30, 37PXX, 14D20, 14D22.

Library of Congress Cataloging-in-Publication Data

Silverman, Joseph H.
Moduli spaces and arithmetic dynamics / Joseph H. Silverman.
p. cm. — (CRM monograph series, Centre de recherches mathématiques, Montréal : v. 30)
Includes bibliographical references and index.
ISBN 978-0-8218-7582-7 (alk. paper)
1. Moduli theory. 2. Analytic spaces. 3. Ergodic theory. 4. Harmonic analysis. I. Title.

QA331.S519 2012
516.3′5—dc23 2011046247

∞ The paper used in this book is acid-free and falls within the guidelines
established to ensure permanence and durability.
This volume was submited to the American Mathematical Society
in camera ready form by the Centre de Recherches Mathématiques.
Visit the AMS home page at http://www.ams.org/

10 9 8 7 6 5 4 3 2 1 17 16 15 14 13 12

Contents

Preface

This monograph is an expanded version of the notes for a series of lectures delivered at a workshop on *Moduli Spaces and the Arithmetic of Dynamical Systems*, Bellairs Research Institute, Barbados, May 2–9, 2010. As such, the level of exposition is uneven, with some results being worked out in detail, while others are merely sketched or have proofs by citation. The goal is to provide an overview, with enough details and pointers to the existing literature, to give the reader an entry into this exciting area of current research. It is the author's hope that this will be useful, especially since at present there are only a small number of books [4, 38, 61, 90, 99, 110] dealing with the arithmetic or algebraic side of dynamical systems. For further reading, the reader might consult the webpage

`http://www.math.brown.edu/~jhs/ADSHome.html`

which contains links to an extensive list of articles in this area.

Acknowledgements. I would like to thank Xander Faber and the McGill University Mathematics Department for inviting me to deliver a series of lectures at the Bellairs Workshop, and Chantal David and Ina Mette for arranging for these notes to be published as a CRM monograph.

I would like to thank Shu Kawaguchi for showing me the proof of Proposition 3.18, which he adapted from [52, Proposition 21.6], Michelle Manes for providing the proof sketch of Bousch's theorem (Theorem 4.12), Michelle Manes and Alon Levy for the content of Remark 4.19, Tom Scanlon for providing the proof sketch of Theorem 5.11(b), Adam Epstein for providing the proof sketch of Proposition 6.18, Michael Zieve for allowing me to include the algebraic characterization of Lattès maps (Theorem 3.22), Curt McMullen for providing information about transcendence in dynamics, and Xavier Buff, Laura DeMarco, and Adam Epstein for a number of helpful email conversations.

I would also like to thank the people who looked at an initial draft of this monograph and offered suggestions and corrections: Adam Epstein, Ben Hutz, Patrick Ingram, Michelle Manes, Bjorn Poonen.

These notes greatly benefited from the many questions and comments posed by the participants at the Bellairs Workshop, so I would like to thank all of them for being such a lively audience: Arthur Baragar, Anupam Bhatnagar, Henri Darmon, Laura DeMarco, Adam Epstein, Xander Faber, Eyal Goren, Benjamin Hutz, Patrick Ingram, Rafe Jones, Shu Kawaguchi, Cristin Kenney, Sarah Koch, Holly Krieger, ChongGyu (Joey) Lee, Alon Levy, Kalyani Madhu, Michelle Manes, Alice Medvedev, Bjorn Poonen, Michael L. Tepper, Adam Towsley, and Phillip Williams.

Joseph H. Silverman
August 2011

Introduction

At its most abstract, discrete dynamics is the study of iteration of a self-map $\phi\colon S \to S$ of a set. Even in this minimalist context, it is often an interesting problem to characterize the points of S according to whether their orbits are infinite or finite, and among the latter, whether the orbit is purely periodic. Many other questions may be generated by adding additional structure to S, for example of a topological, analytic, or algebraic nature, and requiring that the map ϕ respect this structure. So for example, if S is a topological space and ϕ is continuous, we might ask which points in S have a dense orbit, while if S is equipped with a measure, we can study equidistribution properties of orbits. And to take two specific spaces of considerable interest, the dynamics of polynomial and holomorphic self-maps of $\mathbb{C}$ and rational maps of $\mathbb{P}^1(\mathbb{C})$ have spawned a vast literature that includes many deep and fascinating results.

A meta-mathematical principle is that first one studies (isomorphism classes of) objects, then one studies maps between objects that preserve the objects' properties, then the maps themselves become objects for study and one tries to put a "nice" structure on the collection of maps (often modulo some sort of equivalence relation).[1]

In many cases of interest the set of maps can be given the structure of a manifold or algebraic variety. Denoting the set of maps by $\mathcal{P}$ (for parameter space), the relation that identifies maps as being equivalent is often given by the action of a group G on the parameter space $\mathcal{P}$. It is then important, and useful, to understand what structure can be given to the quotient space $\mathcal{P}/G$, which we denote by $\mathcal{M}$ (for moduli space). For example, can we give $\mathcal{M}$ the structure of a manifold or orbifold or variety or scheme or stack or...? Further, if $\mathcal{M}$ has such a structure, it is generally not a complete (compact) space, so we will want to find a completion $\overline{\mathcal{M}}$ and to describe how the points in the boundary $\overline{\mathcal{M}} \smallsetminus \mathcal{M}$ correspond to degenerate versions of the maps classified by $\mathcal{M}$.

This, in brief, is the topic of this monograph. We will focus almost exclusively on the collection of holomorphic (or sometimes meromorphic) self-maps of $\mathbb{P}^n(\mathbb{C})$, or more generally morphisms (or rational) self-maps of $\mathbb{P}^n$ over an arbitrary field K. The natural equivalence relation to impose is that $\phi \sim \psi$ if there is an automorphism $f \in \operatorname{Aut}(\mathbb{P}^n) = \mathrm{PGL}_{n+1}$ such that $\psi = f^{-1} \circ \phi \circ f$, since this conjugation action commutes with composition of functions and corresponds to a simultaneous change-of-variables on the domain and range. The quotient space of self-maps of $\mathbb{P}^n$ of degree d modulo PGL-conjugation is denoted M_d^n.

We begin in Chapter 1 with an overview of dynamical systems and their parameter and moduli spaces. Chapter 2 is devoted to the construction of dynamical

[1]We'll stop here, but one can continue this process and study maps on space(s) of maps, etc.

moduli spaces M_d^n via geometric invariant theory (GIT). For the convenience of the reader, we include a brief introduction (without proofs) to the parts of GIT that are used in the construction of M_d^n. Chapter 2 also includes a description of some basic geometric properties of M_d^1. Chapter 3 covers further topics related to dynamical moduli spaces. In Chapter 4 we restrict attention to the dynamical modular curves whose points classify pairs (ϕ_c, α), where ϕ_c is the quadratic polynomial $z^2 + c$ and α is a ϕ_c-periodic point of some specified period. Chapter 5 introduces canonical heights, which are used to study both the arithmetic complexity of points relative to maps and the maps themselves. In Chapter 6 we discuss maps that are postcritically finite, i.e., all of whose critical points have finite forward orbit. The dynamics of these maps has been much studied, and we examine the locus in moduli space of such maps. Finally, Chapter 7 examines the distinction between $\mathrm{PGL}_{n+1}(\overline{K})$-equivalence and $\mathrm{PGL}_{n+1}(K)$-equivalence. The former suffices for studying geometry, but the latter is important when studying number theoretic properties of dynamical systems.

CHAPTER 1

Moduli Spaces Associated to Dynamical Systems

In this chapter we give an introduction to the objects that we will study, dynamical systems, abstract moduli spaces, dynamical parameter spaces, and dynamical moduli spaces. Ultimately we will want to study both the geometry and the arithmetic of these objects. A basic reference for arithmetic dynamics is [110], and we refer the reader especially to Chapter 4 of [110], which deals with moduli questions.

1.1. Dynamical definitions

Discrete dynamics is the study of iteration of functions. Abstractly, we start with a set X (generally having some additional structure) and a self-map (satisfying appropriate properties)

$$\phi\colon X \longrightarrow X$$

from X to itself. The map ϕ may be iterated to yield new self-maps of X; we write

$$\phi^n = \underbrace{\phi \circ \phi \circ \cdots \circ \phi}_{n \text{ iterations}}$$

for the nth iterate of ϕ.

The iterates of ϕ applied to a point $x \in X$ give the *(forward) orbit of* x, which we denote by

$$\mathcal{O}_\phi(x) = \{x, \phi(x), \phi^2(x), \phi^3(x), \dots\}.$$

A fundamental problem in discrete dynamics is the study of orbits, and in particular, the classification of points in X according to the behavior of their orbits.

Example 1.1. We consider dynamical systems defined by analytic self-maps of complex manifolds. For example, taking $X = \mathbb{C}$, we could look at a polynomial map $\phi(z) \in \mathbb{C}[z]$ or at the exponential map $\phi(z) = e^z$. The dynamics of polynomial maps differ in many ways from the dynamics of the map e^z.

Suppose instead that we study analytic dynamics on the compact manifold $X = \mathbb{P}^1(\mathbb{C})$. Aside from the constant map $\phi(z) = \infty$, the only analytic self-maps of $\mathbb{P}^1(\mathbb{C})$ are rational functions $\phi(z) \in \mathbb{C}(z)$. We observe that the set of self-maps of $\mathbb{P}^1(\mathbb{C})$, which we denote by $\operatorname{Hom}(\mathbb{P}^1(\mathbb{C}))$, breaks up into a disjoint countable union consisting of the sets

$$\operatorname{Hom}_d(\mathbb{P}^1(\mathbb{C})) = \{\phi \in \operatorname{Hom}(\mathbb{P}^1(\mathbb{C})) : \deg(\phi) = d\} \quad \text{for } d = 0, 1, 2, \dots.$$

Further, as we will see, $\operatorname{Hom}_d(\mathbb{P}^1(\mathbb{C}))$ has a natural structure as a Zariski open subset of $\mathbb{P}^{2d+1}(\mathbb{C})$, so for each $d \geq 0$, there is a continuously varying $(2d+1)$-dimensional family of rational maps of degree d.

In fancy terminology, we start with a category $\mathcal{C}$ whose objects are sets. Then each object $X \in \operatorname{Obj}(\mathcal{C})$ and each self-morphism $\phi \in \operatorname{Hom}(X, X)$ yields a dynamical system (X, ϕ), since we can iterate the map ϕ and apply it to the points of X.

For convenience, we write $\operatorname{Hom}(X)$ for the self-morphisms of X. (The standard categorical notation would be $\operatorname{Hom}(X,X)$ or $\operatorname{End}(X)$.)

Definition 1.2. Two self-morphisms $\phi,\psi\in\operatorname{Hom}(X)$ are *equivalent* (or *conjugate*) if there is an automorphism, i.e., a self-isomorphism $f\in\operatorname{Aut}(X)$ that makes the following square commute:

$$\begin{array}{ccc} X & \xrightarrow{\ \phi\ } & X \\ \big\downarrow f & & \big\downarrow f \\ X & \xrightarrow{\ \psi\ } & X. \end{array}$$

We write

$$\phi^f = f^{-1}\circ\phi\circ f$$

for the conjugation of ϕ by f, so ψ is equivalent to ϕ if $\psi=\phi^f$ for some $f\in\operatorname{Aut}(X)$. We may view conjugation as "changing coordinates" on X. This makes it the "right" equivalence relation for dynamics, since simultaneously making the same change of coordinates on the domain and range of ϕ gives a relation that commutes with iteration,

$$(\phi^f)^n = (\phi^n)^f.$$

Further, it is easy to see that conjugation transforms orbits in a natural way,

$$\mathcal{O}_{\phi^f}(x) = f\left(\mathcal{O}_\phi(f^{-1}(x))\right).$$

Even if the space X has no further structure, we can still classify orbits according to their length.

Definition 1.3. A point $x\in X$ if *periodic for ϕ* if

$$\phi^n(x)=x \quad \text{for some } n\ge 1.$$

We write $\operatorname{Per}(\phi,X)$ for the set of ϕ-periodic points, and we let

$$\operatorname{Per}_n(\phi,X)=\bigl\{x\in X:\phi^n(x)=x\bigr\}.$$

The (*exact*) *period* of x is the smallest n such that $\phi^n(x)=x$. A point of period one is called a *fixed point.* There is also a notion of *formal period*, which we discuss in Section 4.1.[1]

A point $x\in X$ if *preperiodic for ϕ* if its orbit is finite, or equivalently if some point in its orbit is periodic. We write $\operatorname{PrePer}(\phi,X)$ for the set of ϕ-preperiodic points, and we let

$$\operatorname{PrePer}_{m,n}(\phi,X)=\bigl\{x\in X:\phi^{m+n}(x)=\phi^m(x)\bigr\}.$$

In particular, $\operatorname{PrePer}_{0,n}(\phi,X)=\operatorname{Per}_n(\phi,X)$.

When the space X is fixed, we omit it from the notation and write $\operatorname{Per}(\phi)$ and $\operatorname{PrePer}(\phi)$.

Remark 1.4. If $\phi,\psi\in\operatorname{Hom}(X)$ satisfy

$$f\circ\phi=\psi\circ f \tag{1.1}$$

[1] In classical dynamical terminology, the exact period is called the *prime period*, which we deprecate because of possible confusion with periods that are prime numbers.

for some morphism $f \in \operatorname{Hom}(X)$ that need not be an automorphism, then ϕ and ψ are said to be *semiconjugate*. Semi-conjugation is also useful in studying dynamics, since (1.1) implies that

$$f \circ \phi^n = \psi^n \circ f \quad \text{for all } n \geq 0.$$

1.2. Moduli spaces: what they are and why they're useful

We start with an abstract formulation of a *moduli problem*. Suppose that we are given:

- A collection of objects parametrized by a parameter space $\mathcal{P}$.
- An equivalence relation $\sim$ on $\mathcal{P}$.

The goal is to classify the equivalence classes in a functorial way. A *moduli space* for this problem is a space $\mathcal{M}$ that is "functorially" isomorphic to the quotient $\mathcal{P}/{\sim}$. We illustrate with a familiar example.

Example 1.5. Consider

$$\mathcal{P} = \bigl\{(A,B) : 4A^3 + 27B^2 \neq 0\bigr\}.$$

Then $\mathcal{P}$ parametrizes elliptic curves given by Weierstrass equations[2]

$$(A,B) \in \mathcal{P} \;\rightsquigarrow\; E_{A,B} : Y^2 = X^3 + AX + B.$$

Our equivalence relation is isomorphism of the associated elliptic curves, i.e.,

$$\begin{aligned}(A,B) \sim (A',B') &\iff E_{A,B} \text{ is isomorphic to } E_{A',B'} \\ &\iff A = u^4 A' \text{ and } B = u^6 B' \text{ for some } u \neq 0.\end{aligned}$$

The moduli space $\mathcal{M} = \mathcal{P}/{\sim}$ may be identified with the affine line $\mathbb{A}^1$ via the j-invariant map

$$j \colon \mathcal{P} \longrightarrow \mathbb{A}^1, \quad j(A,B) = \frac{4A^3}{4A^3 + 27B^2}.$$

We now describe the key functorially property that makes $\mathbb{A}^1$ the moduli space for this moduli problem. Let

$$E \longrightarrow T$$

be any algebraic family of elliptic curves parametrized by the points of a variety T. Then the natural map of sets

$$T \longrightarrow \mathbb{A}^1, \quad t \longmapsto j(E_t),$$

is automatically a *morphism* of varieties.

1.3. Fine moduli spaces and coarse moduli spaces

Suppose now that the parameter space $\mathcal{P}$ is a variety (scheme, stack,...) that parametrizes some sort of algebro-geometric object, e.g., curves of genus g or principally polarized abelian varieties of dimension d, and that $\sim$ identifies objects that are isomorphic.

A (*coarse*) *moduli space* is a variety (scheme, stack,...) $\mathcal{M}$ with the following properties:

- $\mathcal{M}(k) = \mathcal{P}(k)/{\sim}$ for all algebraically closed fields.

[2]To simplify the exposition, we are being a little imprecise. What we should say is that $\mathcal{P}$ parametrizes elliptic curves defined over an algebraically closed field of characteristic not equal to 2 or 3.

- If $X \to T$ is an algebraic family of the type of objects parametrized by $\mathcal{P}$, then the induced map of sets,

$$T \longrightarrow \mathcal{M}, \quad t \longmapsto (\text{equivalence class of } X_t), \tag{1.2}$$

is a morphism.

We might ask for more, namely we might want a *universal family*. A *fine moduli space* is a pair of varieties (schemes, stacks,...) $(\mathcal{X}, \mathcal{M})$ and a map

$$\mathcal{X} \to \mathcal{M}$$

such that:

- $\mathcal{M}$ is a coarse moduli space.
- For each $m \in \mathcal{M}$, the fiber $\mathcal{X}_m$ is in the isomorphism class classified by m.
- For any algebraic family $X \to T$ of objects, let $f : T \to \mathcal{M}$ be the map (1.2) coming from the fact that $\mathcal{M}$ is a course moduli space. Then there are morphisms making a commutative square

$$\begin{array}{ccc} X & \longrightarrow & \mathcal{X} \\ \downarrow & & \downarrow \\ T & \xrightarrow{f} & \mathcal{M} \end{array}$$

such that the restriction to every fiber $X_t \to \mathcal{X}_{f(t)}$ is an isomorphism.

Example 1.6. There is no fine moduli space for the elliptic curve problem. If there were such a space $\mathcal{E} \to \mathbb{A}^1$, then as you walk around the $j = 0$ or $j = 1728$ points, the fiber would be subject to a nontrivial automorphism. And even ignoring these special points, the fact that elliptic curves have an automorphism of order two, i.e., $P \to -P$, precludes the existence of a fine moduli space.

This is a general phenomenon. If $\mathcal{M}$ is a coarse moduli space and if some of the objects admit nontrivial automorphisms, then there is no universal family. A standard way to get rid of automorphisms is to add level structure, thereby allowing the construction of a fine moduli space for objects with level structure.

Example 1.7. The moduli problem

$$\left\{ \begin{array}{l} \text{isomorphism classes of elliptic curves} \\ \text{with a marked point of order } N \end{array} \right\}$$

admits a fine moduli space provided $N > 3$, since an automorphism of an elliptic curve never fixes any point of order $N > 3$.

1.4. Parameter spaces for dynamical systems

We begin with dynamical systems on the projective line $\mathbb{P}^1$. In the analytic or algebraic categories, the self-maps of $\mathbb{P}^1$ are rational maps, which leads to the dynamical parameter spaces

$$\mathrm{Hom}_d = \{\text{degree } d \text{ rational maps } \mathbb{P}^1 \to \mathbb{P}^1\}.$$

An element $\phi \in \mathrm{Hom}_d$ is a rational function

$$\phi(z) = \frac{a_0 z^d + a_1 z^{d-1} + \cdots + a_{d-1} z + a_d}{b_0 z^d + b_1 z^{d-1} + \cdots + b_{d-1} z + b_d} = \frac{F_a(z)}{F_b(z)}$$

with at least one of a_d and b_d not zero and with $\gcd(F_a, F_b) = 1$. The map ϕ does not change if we multiply the numerator and the denominator by a nonzero number,

$$\phi(z) = \frac{\lambda F_a(z)}{\lambda F_b(z)} \quad \text{for any } \lambda \neq 0.$$

Thus Hom_d has a natural description as a subvariety of projective space,

$$\begin{aligned} \mathrm{Hom}_d &= \bigl\{[a,b] \in \mathbb{P}^{2d+1} : a_0, b_0 \text{ not both } 0 \text{ and } \gcd(F_a, F_b) = 1\bigr\} \\ &= \bigl\{[a,b] \in \mathbb{P}^{2d+1} : \mathrm{Resultant}(F_a, F_b) \neq 0\bigr\}. \end{aligned}$$

Every rational self-map $\mathbb{P}^1 \to \mathbb{P}^1$ is automatically a morphism, but this is no longer true in higher dimension. We let

$$\begin{aligned} \mathrm{Rat}_d^n &= \{\text{degree } d \text{ rational maps } \mathbb{P}^n \dashrightarrow \mathbb{P}^n\}, \\ \mathrm{Hom}_d^n &= \{\text{degree } d \text{ morphisms } \mathbb{P}^n \to \mathbb{P}^n\}. \end{aligned}$$

Note that maps in Hom_d^n can be iterated, but iterates of general rational maps need not be well-defined. However, dominant rational maps can be iterated, although some points may have orbits that terminate in the indeterminacy locus of the map.

Every $\phi \in \mathrm{Rat}_d^n$ has the form

$$\phi = [\phi_0, \ldots, \phi_n],$$

where each ϕ_i is a homogeneous polynomial of degree d in $X_0, \ldots, X_n$, and the polynomials $\phi_0, \ldots, \phi_n$ have no common factor. The map ϕ is in Hom_d^n if and only if $\phi_0, \ldots, \phi_n$ have no nontrivial common zeros. Writing

$$\phi_i = \sum_j a_{ij} X^j \quad \text{using a multi-index } j = (j_0, \ldots, j_n),$$

we identify ϕ with a point in projective space

$$\phi \leftrightarrow [a_{ij}]_{i,j} \in \mathbb{P}^N, \quad \text{where } N = N(n,d) = \binom{n+d}{d}(n+1) - 1. \tag{1.3}$$

Conversely, every point in $\mathbb{P}^N$ is identified with a rational map $\mathbb{P}^n \dashrightarrow \mathbb{P}^n$ of some degree. We observe that

$$\mathrm{Hom}_d^n \subset \mathrm{Rat}_d^n \subset \mathbb{P}^N$$

are Zariski open subsets of $\mathbb{P}^N$. For Hom_d^n we have the following precise result.

Theorem 1.8. *There exists a geometrically irreducible polynomial $\mathcal{R} \in \mathbb{Z}[a_{ij}]$ in the coefficients of ϕ such that*

$$\phi \in \mathrm{Hom}_d^n \iff \mathcal{R}(\phi) \neq 0.$$

In particular, Hom_d^n is an affine variety. The polynomial $\mathcal{R}$ is called the Macaulay resultant *of ϕ. It is multihomogeneous in the coefficients of the coordinate polynomials $\phi_0, \ldots, \phi_n$ defining ϕ.*

PROOF. See [25, Theorem 3.8], [29, Chapter 3], or [56]. □

Remark 1.9. For a map $\phi = [\phi_0, \ldots, \phi_n] \in \mathrm{Hom}_d^n$, the homogeneous polynomials $\phi_0, \ldots, \phi_n$ are only determined up to multiplication by a nonzero constant c. The Macaulay resultant $\mathcal{R}$ depends on the choice of $\phi_0, \ldots, \phi_n$, so is only well-defined up to a power of c; more precisely,

$$\mathcal{R}(c\phi_0, \ldots, c\phi_n) = c^{(n+1)d^n}\, \mathcal{R}(\phi_0, \ldots, \phi_n).$$

In any case, the question of whether $\mathcal{R}(\phi)$ vanishes is independent of the choice of coordinate functions for ϕ, so Theorem 1.8 makes sense. We will use the Macaulay resultant later to define the minimal resultant of a rational map; see Section 3.2.

1.5. Moduli spaces for dynamical systems

The projective linear group

$$\mathrm{PGL}_{n+1} = \mathrm{Aut}(\mathbb{P}^n)$$

acts on Rat_d^n and Hom_d^n in various ways. For dynamics we are interested in the conjugation action:

$$\phi^f = f^{-1} \circ \phi \circ f \quad \text{for } \phi \in \mathrm{Rat}_d^n \text{ and } f \in \mathrm{PGL}_{n+1}.$$

As noted earlier, this action satisfies

$$(\phi^n)^f = (\phi^f)^n$$

and corresponds to change of variables for iteration:

$$\begin{array}{ccc} \mathbb{P}^n & \xrightarrow{\ \phi\ } & \mathbb{P}^n \\ \big\downarrow f & & \big\downarrow f \\ \mathbb{P}^n & \xrightarrow{\ \phi^f\ } & \mathbb{P}^n. \end{array}$$

The *moduli space of self-morphisms of* $\mathbb{P}^n$ is the set

$$\mathsf{M}_d^n = \mathrm{Hom}_d^n / \mathrm{PGL}_{n+1}.$$

Its points classify self-maps of $\mathbb{P}^n$ up to PGL_{n+1}-conjugation. For now, M_d^n is just a set. Our primary goal in Chapter 2 will be to give M_d^n an algebraic structure. Since the dimension one case comes up so frequently, to ease notation when $n = 1$ we write

$$\mathsf{M}_d = \mathsf{M}_d^1.$$

1.6. Level structure and the uniform boundedness conjecture

It is often useful to look at parameter and moduli spaces that classify a map together with one or more marked periodic points. So for example,

$$\mathrm{Hom}_d^n(m) = \bigl\{(\phi, P) : \phi \in \mathrm{Hom}_d^n \text{ and } P \in \mathrm{Per}_m^{**}(\phi)\bigr\},$$

where $\mathrm{Per}_m^{**}(\phi)$ denotes the set of periodic points of exact period m.[3] Notice that PGL_{n+1} acts on $\mathrm{Hom}_d^n(m)$ via

$$(\phi, P)^f = \bigl(\phi^f, f^{-1}(P)\bigr),$$

so we can form the quotient

$$\mathsf{M}_d^n(m) = \mathrm{Hom}_d^n(m) / \mathrm{PGL}_{n+1}.$$

Study of rational and algebraic points in $\mathsf{M}_d^n(m)$ is closely related to the following motivating conjecture in arithmetic dynamics.

[3]The set $\mathrm{Per}_m^{**}(\phi)$ of points of exact period m is not quite the right set. Instead one uses points of *formal period* m, which may include some points of lower exact period; cf. Section 4.1.

Conjecture 1.10 (Uniform Boundededness Conjecture; Morton–Silverman [84]). *For all n, d, D there is a constant $C(n, d, D)$ such that for all number fields $K/\mathbb{Q}$ of degree at most D and all rational maps $\phi \in \operatorname{Hom}_d^n(K)$,*

$$\#\operatorname{PrePer}\bigl(\phi, \mathbb{P}^n(K)\bigr) \le C(n, d, D).$$

The simplest nontrivial case of the conjecture is for quadratic polynomials

$$\phi_c(z) = z^2 + c \quad \text{with } K = \mathbb{Q}.$$

Let

$$\begin{aligned} \mathcal{Q}_m &= \bigl\{c \in \mathbb{Q} : \operatorname{Per}_m^{**}(\phi_c, \mathbb{Q}) \ne \varnothing\bigr\} \\ &= \{c \in \mathbb{Q} : \phi_c \text{ has a } \mathbb{Q}\text{-rational periodic point of exact period } m\}. \end{aligned}$$

Table 1.1 describes the meager results currently known about $\mathcal{Q}_m$. Poonen conjectures that $\mathcal{Q}_m = \varnothing$ for all $n \ge 4$.

To illustrate the depth of the uniform boundedness conjecture, we observe that it trivially implies Mazur's uniform boundedness theorem [72] for torsion on elliptic curves over $\mathbb{Q}$, and similarly Merel's theorem [77] for number fields.

Proposition 1.11. *Suppose that the uniform boundedness conjecture is true for $(n, d, D) = (1, 4, 1)$, i.e., for maps $\mathbb{P}^1 \to \mathbb{P}^1$ of degree 4 defined over $\mathbb{Q}$. Then*

$$\#E(\mathbb{Q})_{\mathrm{tors}} \le C(1, 4, 1) \quad \textit{for all elliptic curves } E/\mathbb{Q},$$

where $C(1, 4, 1)$ is the constant appearing in Conjecture 1.10.

PROOF. Consider the Lattès map ϕ_E defined by the commutativity of the diagram

$$\begin{array}{ccc} E & \xrightarrow{P \mapsto 2P} & E \\ \big\downarrow & & \big\downarrow \\ \mathbb{P}^1 & \xrightarrow{\ \phi_E\ } & \mathbb{P}^1. \end{array}$$

Then

$$E(\mathbb{Q})_{\mathrm{tors}} = \operatorname{PrePer}\bigl(\phi_E, \mathbb{P}^1(\mathbb{Q})\bigr).$$

so $\#E(\mathbb{Q})_{\mathrm{tors}} \le C(1, 4, 1)$. □

Although not so immediate, the full uniform boundedness conjecture implies an analogous result for torsion points on abelian varieties. This latter statement is not currently known unconditionally for any number field in any dimension greater than one.

TABLE 1.1. $\mathcal{Q}_m = \bigl\{c \in \mathbb{Q} : \operatorname{Per}_m^{**}(\phi_c, \mathbb{Q}) \ne \varnothing\bigr\}$

m	$\#\mathcal{Q}_m$	Notes
1,2,3	∞	Morton [82]
4	0	Morton [82]
5	0	Flynn, Poonen, Schaefer [42]
6	0	Stoll [115] (assuming B–SwD)
≥ 7	???	

Theorem 1.12 (Fakhruddin [39,40]**).** *Assume that the uniform boundedness conjecture is true. Then for every number field K and abelian variety A/K we have*

$$\# A(K)_{\mathrm{tors}} < C\big(\dim A, [K:\mathbb{Q}]\big),$$

where as indicated the constant C depends only on the dimension of the abelian variety A and the degree of the number field K.

PROOF SKETCH. The idea of the proof is to embed A into projective space in such a way that the duplication map

$$A \xrightarrow{[2]} A \tag{1.4}$$

extends to a morphism of the projective space. More generally, Fakhruddin shows the following. Let $\phi\colon X \to X$ is a polarized self-morphism of a projective variety, i.e., there is an ample divisor $D \in \mathrm{Div}(X)$ and an integer $n \geq 2$ such that $\phi^* D \sim nD$. Then there is a projective embedding $X \hookrightarrow \mathbb{P}^N$ such that ϕ extends to a self-morphism of $\mathbb{P}^N$. Further, this can be done functorially, so for abelian varieties of a fixed dimension and for the duplication map (which has fixed degree), the associated projective space and self-morphism also have fixed dimension and degree. We also mention that for the map (1.4), one might be able to do explicitly write down Fakhruddin's map using Mumford's theory of equations defining abelian varieties [85–87]. □

CHAPTER 2

The Geometry of Dynamical Moduli Spaces

The group PGL_{n+1} acts on Hom_d^n via conjugation, and this action extends in a natural way to all of $\mathbb{P}^N$, where we embed

$$\mathrm{Hom}_d^n \hookrightarrow \mathbb{P}^N$$

as described by (1.3) on page 7. Our goal is to give the quotient

$$\mathrm{Hom}_d^n \,/\, \mathrm{PGL}_{n+1}$$

the structure of a variety (scheme, stack,...) and study its geometry (affine, projective, nonsingular, rational,...). More ambitiously, we ask for the largest subset of $\mathbb{P}^N$ whose quotient by PGL_{n+1} admits a nice structure. For example, it would be nice if this were true of the space of degree d rational maps Rat_d^n, but unfortunately if $n \geq 2$, then Rat_d^n is too large.

We begin with an easier case. Let

$$V \text{ be a variety} \quad \text{and let} \quad G \subset \mathrm{Aut}(V) \text{ be a } \textit{finite} \text{ group.}$$

Then it is not hard to construct a variety that plays the role of the quotient V/G. It's enough to do it for affine varieties and then glue them together. For an affine variety, we have

$$V = \operatorname{Spec} A, \quad \text{and we define} \quad V/G \stackrel{\text{def}}{=} \operatorname{Spec} A^G,$$

where A^G is the ring of invariants,

$$A^G = \bigl\{a \in A : g^*(a) = a \text{ for all } g \in G\bigr\}.$$

(Note that $g \in \mathrm{Aut}(V) = \mathrm{Aut}(\operatorname{Spec} A)$, so $g : V \to V$ induces a ring homomorphism $g^* \colon A \to A$.) The fact that A^G is a finitely generated algebra, and hence that $\operatorname{Spec} A^G$ is a variety, is a consequence of Hilbert's theorem.

Example 2.1. We consider the permutation action of $\mathbb{Z}/2\mathbb{Z}$ on the coordinates of the affine plane:

$$V = \mathbb{A}^2 = \operatorname{Spec} k[x,y], \quad G = \mathbb{Z}/2\mathbb{Z} = \{1, \sigma\}, \quad \sigma(x) = y, \quad \sigma(y) = x.$$

Then

$$V/G = \operatorname{Spec} k[x,y]^G = \operatorname{Spec} k[x+y, xy] \cong \mathbb{A}^2.$$

More generally, let the symmetric group $\mathcal{S}_n$ act on $\mathbb{A}^n = \operatorname{Spec} k[x_1,\ldots,x_n]$ by permuting the coordinates. Then a classical theorem on symmetric polynomials says that

$$\mathbb{A}^n/\mathcal{S}_n = \operatorname{Spec} k[x_1,\ldots,x_n]^{\mathcal{S}_n} = \operatorname{Spec} k[s_1,\ldots,s_n] \cong \mathbb{A}^n,$$

where s_i is ith elementary symmetric polynomial in $x_1,\ldots,x_n$.

This handles finite groups, but if G is an infinite group, then even for "nice" actions, the quotient V/G often doesn't exist as a "nice" object. Mumford invented *Geometric Invariant Theory* (GIT) to study this problem. The name comes from the fact that A^G is a *ring of invariants.* The problem of computing and describing rings of invariants has been extensively studied since the 19th century. A notable example is Hilbert's theorem, already alluded to above, which says that if A is a finitely generated algebra over a field k and if G is finite, then A^G is also a finitely generated k-algebra.

2.1. Introduction to geometric invariant theory (GIT)

The goal of this section is to describe how GIT works, and then to use GIT to construct the quotient space $\operatorname{Hom}_d^n / \operatorname{PGL}_{n+1}$ and some related larger spaces. We refer the reader to [88] for the proofs of the main theorems of GIT; for other expositions of GIT, see for example [32] and [89]. In addition, to keep the exposition manageable, we will generally work with

Varieties over Algebraically Closed Fields,

although much of what we say applies to arbitrary fields, and indeed to schemes over $\operatorname{Spec}\mathbb{Z}$.

Our basic setup consists of the following:

$$\begin{aligned} X/k &\quad \text{a variety,} \\ G/k &\quad \text{an algebraic group,} \\ \sigma\colon G\times X\to X &\quad \text{an algebraic group action.} \end{aligned}$$

In other words, the map σ is a morphism of algebraic varieties that satisfies the usual group action axioms,

$$\sigma(1,x)=x \quad\text{and}\quad \sigma\bigl(g_1,\sigma(g_2,x)\bigr)=\sigma(g_1g_2,x).$$

For convenience, we will often write $g\cdot x$ as an abbreviation for $\sigma(g,x)$.

We observe that any morphism

$$f\colon T\longrightarrow X$$

induces a morphism

$$\psi_f\colon G\times T\longrightarrow X\times T,\quad (g,t)\longmapsto \bigl(g\cdot f(t),t\bigr).$$

In other words, there is a functorial action of G on the T-valued points $X(T)$ of X. For example, if $T=X$ and $f=1_X$, then

$$\psi_f\colon G\times X\to X\times X,\quad \psi_f(g,x)=(g\cdot x,x).$$

Definition 2.2. Let $f\in X(T)$. The *orbit of f* is the set

$$\mathcal{O}_G(f)=\operatorname{Image}(\psi_f)\subset X\times T.$$

In terms of points,

$$\mathcal{O}_G(f)=\bigl\{(g\cdot f(t),t)\in X\times T : g\in G, t\in T\bigr\}.$$

If $x\in X$, we let $f_x\colon\{x\}\hookrightarrow X$ and write

$$\mathcal{O}(x)\text{ or }\mathcal{O}_G(x)\quad\text{for } p_1\bigl(\mathcal{O}(f_x)\bigr)\subset X.$$

Definition 2.3. Let $f \in X(T)$. The *stabilizer of f* is the fiber product

$$\mathrm{Stab}(f) = (G \times T) \times_{X \times T} T,$$

where the fibering is via the maps ψ_f and $f \times 1_T$. In terms of points, this is

$$\mathrm{Stab}(f) = \bigl\{(g,t) \in G \times T : g \cdot f(t) = f(t)\bigr\}.$$

We would like to endow X/G with additional structure while retaining the natural quotient properties. For example, at a minimum we want the quotient X/G to have the following properties.

Definition 2.4. A *categorical quotient* for X/G is a variety Y and a morphism

$$\phi\colon X \to Y$$

such that there is a commutative square

$$\begin{array}{ccc} G \times X & \xrightarrow{\ \sigma\ } & X \\ \Big\downarrow{\scriptstyle p_2} & & \Big\downarrow{\scriptstyle \phi} \\ X & \xrightarrow{\ \phi\ } & Y \end{array} \tag{2.1}$$

and such that any variety Z fitting into such a square admits a unique commuting map $Y \to Z$. In other words, if $\psi\colon X \to Z$ is a morphism satisfying $\psi \circ p_2 = \psi \circ \sigma$, then there is a unique morphism $\lambda\colon Y \to Z$ satisfying $\psi = \lambda \circ \phi$.

Note that at the level of points, the commutative square (2.1) agrees with the intuition that we want

$$\begin{array}{ccc} \phi(x) & = & \phi(g \cdot x) \\ \| & & \| \\ \phi \circ p_2(g,x) & & \phi \circ \sigma(g,x) \end{array} \qquad \text{for all } x \in X \text{ and all } g \in G.$$

The definition of categorical quotient is natural, and categorical quotients have many nice properties, such as in the following proposition.

Proposition 2.5. *Let Y be a categorical quotient of X/G. Then*

$$\begin{aligned} X \textit{ reduced} &\implies Y \textit{ reduced.} \\ X \textit{ connected} &\implies Y \textit{ connected.} \\ X \textit{ irreducible} &\implies Y \textit{ irreducible.} \end{aligned}$$

PROOF. See [88, Chapter 0, Remark (2)]. □

However, categorical quotients also lack some properties that we might feel that a quotient should possess. In particular, it seems natural that two points $x, x' \in X$ should have the same image in the quotient if and only if $x' = g \cdot x$ for some $g \in G$. Unfortunately, categorical quotients do not necessarily have this property. Instead, we only have

$$\phi(x) = \phi(x') \text{ in } X/G \iff \overline{\mathcal{O}_G(x)} \cap \overline{\mathcal{O}_G(x')} \neq \varnothing,$$

where $\overline{\mathcal{O}_G(x)}$ denotes the Zariski closure of the orbit of x. Note that an orbit $\mathcal{O}_G(x)$ need not be closed in X, because in general the group G need not be complete.

Definition 2.6. A *geometric quotient* for X/G is a categorical quotient

$$\phi\colon X \to Y$$

with the following additional properties:

(1) ϕ is surjective, and for all $x \in X(k)$, the fiber $X_{\phi(x)} \stackrel{\text{def}}{=} \phi^{-1}\bigl(\phi(x)\bigr)$ satisfies

$$X_{\phi(x)} = \mathcal{O}_G(x).$$

In other words, the geometric fibers of ϕ are precisely the orbits of the geometric points of X.

(2) $U \subset Y$ is open if and only if $\phi^{-1}(U) \subset X$ is open.

(3) The functions on Y are the G-invariant functions on X. In other words, we have $\mathcal{O}_Y = (\phi_*\mathcal{O}_X)^G$, so for any open set $U \subset Y$,[1]

$$\mathcal{O}_Y(U) = \bigl\{f \in \phi_*\mathcal{O}_X\bigl(\phi^{-1}(U)\bigr) : f \circ g = f \text{ for all } g \in G\bigr\}.$$

The group G may not be proper over k, for example, G might be $\mathrm{GL}_n(k)$ or $\mathrm{SL}_n(k)$. This can cause problems when we try to form the quotient, leading to the following definition.

Definition 2.7. The action $\sigma\colon G \times X \to X$ is *closed* if $\mathcal{O}(x)$ is closed for all (geometric) points $x \in X$.

We will be interested primarily in actions of *linear algebraic groups*, i.e., subgroups $G \subset \mathrm{GL}_n$, so there is a natural action of G on $\mathbb{A}^n$. However, not all linear algebraic groups lead to nice quotients, so we consider the following three properties.

Definition 2.8. Let $G \subset \mathrm{GL}_n$ be a linear algebraic group.

- G is *reductive* if its radical is a torus. (The radical is the maximal connected solvable normal subgroup.)
- G is *linearly reductive* if it is completely reducible, i.e., whenever the action of G on $\mathbb{A}^n$ leaves some subspace $\mathbb{A}^m \subset \mathbb{A}^n$ invariant, then it leaves a complementary subspace $\mathbb{A}^{n-m}$ invariant.
- G is *geometrically reductive* if whenever $G \to \mathrm{GL}_n(k)$ is a rational representation and $v \in (k^n)^G$ is a nonzero invariant vector, then there exists a nonzero homogeneous invariant polynomial $F \in k[x_1, \dots, x_n]^G$ such that $F(v) \neq 0$. (Intuition: If G is geometrically reductive, then G-invariant functions can be used to separate points.)

Theorem 2.9. *Let G/k be a linear algebraic group.*

$$\begin{aligned} G \text{ is reductive} &\iff G \text{ is geometrically reductive} \\ &\overset{\mathrm{char}(k)=0}{\iff} G \text{ is linearly reductive.} \end{aligned}$$

Theorem 2.10. *Let G/k be a reductive algebraic group acting on an affine variety X.*

(a) *The categorical quotient X/G exists as an affine variety.*

(b) *If the action of G on X is closed, then X/G is a geometric quotient.*

In general, we want to find a "large" subset of X on which G acts "nicely." We do this relative to an (ample) invertible sheaf $\mathcal{L}$ on X. To make things work, we need to linearize the action of G on $\mathcal{L}$.

---

[1]In this definition, $\mathcal{O}_X$ and $\mathcal{O}_Y$ are the structure sheaves on X and Y, not orbits. It should be clear from context when the letter $\mathcal{O}$ denotes a structure sheaf and when it denotes an orbit.

Definition 2.11. Let $\sigma\colon G \times X \to X$ be a group action, and let $\mathcal{L}$ be an invertible sheaf on X. A *G-linearization of $\mathcal{L}$* is an isomorphism

$$\phi\colon \sigma^*\mathcal{L} \xrightarrow{\sim} p_2^*\mathcal{L}$$

satisfying an appropriate cocycle condition. (Essentially, we want to lift the action of G to the line bundle $L \to X$ associated to $\mathcal{L}$.)

Example 2.12. The moduli spaces associated to dynamical systems are quotient spaces of the form

$$\mathrm{Rat}_d^n / \mathrm{PGL}_{n+1},$$

where the conjugation action of PGL_{n+1} on Rat_d^n is via the embedding

$$\mathrm{Rat}_d^n \subset \mathbb{P}^N$$

as described by (1.3). The action of PGL_{n+1} on $\mathbb{P}^N$ may thus be viewed as an embedding

$$\mathrm{PGL}_{n+1} \hookrightarrow \mathrm{PGL}_{N+1}$$

followed by the natural action of PGL_{N+1} on $\mathbb{P}^N$.

We thus seek an ample line bundle on $\mathbb{P}^N$ for which the action of PGL_{N+1} is linearized. The natural choice is the line bundle $\mathcal{O}_{\mathbb{P}^N}(1)$, but unfortunately this does not work, because there is no consistent way to lift the action of an element $g \in \mathrm{PGL}_{N+1}$ to $\mathcal{O}_{\mathbb{P}^N}(1)$. The problem is that g is only defined up to multiplication by a scalar matrix, so for example, it does not act on the global sections $x_0, \ldots, x_N$ that generate $\mathcal{O}_{\mathbb{P}^N}(1)$.

We could instead use the action of GL_{N+1}, but then the stabilizer of a point includes a copy of the multiplicative group $\mathbb{G}_m$, which would cause problems later. Here are two better ways to surmount this difficulty.

- Use the action of SL_{N+1}. This clearly lifts unambiguously to an action on $\mathcal{O}_{\mathbb{P}^N}(1)$. Further, on geometric points the natural homomorphism
$$\mathrm{SL}_{N+1}(k) \longrightarrow \mathrm{PGL}_{N+1}(k)$$
is surjective and has finite kernel, so we get the same geometric points in the quotient $\mathrm{Rat}_d^n / \mathrm{SL}_{n+1}$, and stabilizers of points are only changed by a finite amount.
- Use the line bundle $\mathcal{O}_{\mathbb{P}^N}(N+1)$. If we let $g(x_i)$ denote the usual action of a matrix g on a coordinate function, then we can linearize the action of PGL_{N+1} on $\mathcal{O}_{\mathbb{P}^N}(N+1)$ by setting
$$g * (x_0^{e_0} x_1^{e_1} \cdots x_N^{e_N}) = \frac{g(x_0)^{e_0} g(x_1)^{e_1} \cdots g(x_N)^{e_N}}{\det g}.$$
Note that this action of PGL_{N+1} on $\mathcal{O}_{\mathbb{P}^N}(N+1)$ is well-defined, independent of the lift of g from PGL_{N+1} to GL_{N+1}, since
$$\sum e_i = N+1 \qquad \text{and} \qquad \det(cg) = c^{N+1}g \quad \text{for any } c \in k^*.$$

In these notes we will use the first option to construct dynamical moduli spaces.

Now we come to two key definitions that characterize "good" points, i.e., points where the action of G is well-behaved. In the following definition, the fundamental example to keep in mind is

$$X = \mathbb{P}^N, \quad G \subset \mathrm{SL}_{N+1}, \quad \mathcal{L} = \mathcal{O}_X(1).$$

Definition 2.13. Let G be a reductive algebraic group acting on a variety X, let $x \in X$ be a geometric point, and let $\mathcal{L}$ be a G-linearized invertible sheaf on X.

(a) The point x is *semistable* (with respect to $\mathcal{L}$) if for some $n \geq 1$ there is an invariant section $s \in H^0(X, \mathcal{L}^{\otimes n})^G$ such that

$$s(x) \neq 0 \quad \text{and} \quad X_s \text{ is affine,}$$

where $X_s = \{y \in X : s(y) \neq 0\}$.

The fundamental idea here is that semistable points can be separated using G-invariant functions. Thus for $X = \mathbb{P}^N$ and $\mathcal{L} = \mathcal{O}_{\mathbb{P}^N}(1)$, a point $x \in X$ is semistable if there is a G-invariant homogeneous polynomial that does not vanish at x.

(b) The point x is *stable* if in addition, the action of G on X_s is closed, i.e., the orbit $\mathcal{O}_G(x')$ is closed for all (geometric) points $x' \in X_s$.

(c) The point x is *unstable* if it is not semistable.

We let

$$\begin{aligned} X^{\text{ss}}(\mathcal{L}) &= \{x \in X : x \text{ is semistable with respect to } \mathcal{L}\}, \\ X^{\text{s}}(\mathcal{L}) &= \{x \in X : x \text{ is stable with respect to } \mathcal{L}\}, \\ X^{\text{s}}_{(0)}(\mathcal{L}) &= \{x \in X^{\text{s}}(\mathcal{L}) : \dim \operatorname{Stab}(x) = 0\}. \end{aligned}$$

Our next task is to give an alternative definition of stability that uses the line bundle associated to the invertible sheaf $\mathcal{L}$.

Definition 2.14. Let V be a vector space and let $G \subset \mathrm{GL}(V)$ be a reductive algebraic group acting on V. Then $x \in V$ is

$$\begin{aligned} \text{unstable} &\quad \text{if } 0 \in \overline{\mathcal{O}_G(x)}, \\ \text{semistable} &\quad \text{if } 0 \notin \overline{\mathcal{O}_G(x)}, \\ \text{stable} &\quad \text{if } \mathcal{O}_G(x) \text{ is closed and } \dim \mathcal{O}_G(x) = \dim G. \end{aligned}$$

Here's how the definitions of stability are related. Let

$$X \subset \mathbb{P}^N, \quad \mathcal{L} = \mathcal{O}_{\mathbb{P}^N}(1), \quad G \subset \mathrm{SL}_{N+1},$$

and let $\pi \colon \mathbb{A}^{N+1} \smallsetminus \{0\} \to \mathbb{P}^N$ be the projection map. The action of G lifts to the affine cone

$$\hat{X} = \pi^{-1}(X) \cup \{0\}$$

of X, and we get compatible G-actions on all of the spaces in the diagram

$$\begin{array}{ccccc} \mathbb{A}^{N+1} & \supset & \mathbb{A}^{N+1} \smallsetminus \{0\} & \xrightarrow{\ \pi\ } & \mathbb{P}^N \\ \cup & & \cup & & \cup \\ \hat{X} & \supset & \hat{X} \smallsetminus \{0\} & \xrightarrow{\ \pi\ } & X. \end{array}$$

Then

$$\begin{aligned} X^{\text{s}}_{(0)}\big(\mathcal{O}_X(1)\big) &= \{x \in X : \text{some } \hat{x} \in \pi^{-1}(x) \text{ is stable in } \mathbb{A}^{N+1}\}, \\ X^{\text{ss}}\big(\mathcal{O}_X(1)\big) &= \{x \in X : \text{some } \hat{x} \in \pi^{-1}(x) \text{ is semistable in } \mathbb{A}^{N+1}\}. \end{aligned}$$

(We remark that if one $\hat{x} \in \pi^{-1}(x)$ is stable, then they all are, and similarly for semistable.)

After these preliminaries, we're ready for the fundamental existence results, which say (roughly):

$$X^{\mathrm{ss}}/G \text{ has a categorical quotient.}$$
$$X^{\mathrm{s}}/G \text{ has a geometric quotient.}$$

Theorem 2.15 (Fundamental Theorem of Geometric Invariant Theory). *Let G be a reductive group acting on a projective variety X, and let $\mathcal{L}_X$ be a G-linearized ample invertible sheaf on X.*

(a) *There exists a categorical quotient*
$$\phi\colon X^{\mathrm{ss}}(\mathcal{L}_X) \to X^{\mathrm{ss}}(\mathcal{L}_X)/G = Y^{\mathrm{ss}}.$$
(b) *The map ϕ is affine and universally submersive.*
(c) *Y^{ss} is quasi-projective. More precisely, there is an ample invertible sheaf $\mathcal{L}_{Y^{\mathrm{ss}}}$ on Y^{ss} such that $\phi^*\mathcal{L}_{Y^{\mathrm{ss}}} = \mathcal{L}_{X^{\mathrm{ss}}}^{\otimes n}$ for some $n \geq 1$, where we write $\mathcal{L}_{X^{\mathrm{ss}}}$ for the restriction of $\mathcal{L}_X$ to X^{ss}.*
(d) *There is an open subset $Y^{\mathrm{s}} \subset Y^{\mathrm{ss}}$ such that*
   (i) *$\phi^{-1}(Y^{\mathrm{s}}) = X^{\mathrm{s}}(\mathcal{L})$,*
   (ii) *$X^{\mathrm{s}}(\mathcal{L}_X) \xrightarrow{\phi} Y^{\mathrm{s}}$ is a geometric quotient $X^{\mathrm{s}}(\mathcal{L}_X)/G$.*

PROOF. See [88, Chapter 1, Theorem 1.10]. □

2.2. Tools for computing the stable and semistable loci

We are now left with a difficult problem:

How can we find the stable and semistable loci?

To solve this problem, Mumford uses 1-*parameter subgroups*, which are group homomorphisms
$$\ell\colon \mathbb{G}_m \longrightarrow G.$$
In order to determine if a point $x \in X$ is semistable or stable, we look at the composition
$$\begin{array}{ccc} \mathbb{G}_m \xrightarrow{\ \ell\ } G & \longrightarrow & X \\ \alpha & \longrightarrow & \ell(\alpha)\cdot x \end{array}$$
and study what happens as $\alpha \to 0$, i.e., what happens as α goes to the boundary of $\mathbb{G}_m$. By varying the 1-parameter subgroup ℓ, we obtain information about what happens to $g \cdot x$ as g moves to the boundary of G.

Definition 2.16. For any 1-parameter subgroup $\ell\colon \mathbb{G}_m \to G$ and point $x \in X$, the homomorphism $\alpha \mapsto \ell(\alpha)\cdot x$ extends uniquely to a morphism (which depends on x),
$$\ell\colon \mathbb{A}^1 \longrightarrow X,$$
since we have assumed that X is proper. However, we note that the extended map no longer factors through G, since G need not be proper. The image of 0 is denoted
$$\ell(0)\cdot x$$
and is called the *specialization of $\ell(\alpha)\cdot x$ as $\alpha \to 0$.*

The action of $\mathbb{G}_m$ on X via ℓ fixes $\ell(0)\cdot x$, since for any $\beta \in \mathbb{G}_m$ we have

$$\ell(\beta)\cdot\ell(\alpha)\cdot x = \ell(\beta\alpha)\cdot x \xrightarrow[\alpha\to 0]{} \ell(0)\cdot x.$$

Hence we get an action of $\mathbb{G}_m$ on the stalk

$$\mathcal{L}_{\ell(0)\cdot x} \cong \mathbb{A}^1.$$

(We are using the G-linearization of $\mathcal{L}$ here, since that's how we get an action of G on $\mathcal{L}$.)

We thus have an algebraic homomorphism

$$\chi_\ell\colon \mathbb{G}_m \xrightarrow{\ \ell\ } \operatorname{Aut}(\mathcal{L}_{\ell(0)\cdot x}) = \operatorname{Aut}(\mathbb{A}^1) = \mathbb{G}_m.$$

Every such homomorphism is given by a character, i.e., there is an integer $r\in\mathbb{Z}$ such that

$$\chi_\ell\colon \mathbb{G}_m \longrightarrow \mathbb{G}_m,\quad \chi_\ell(\alpha)=\alpha^r.$$

This allows us to make a important definition.

Definition 2.17. With notation as above, we let

$$\mu^{\mathcal{L}}(x,\ell) = -r,\quad \text{where } \chi_\ell(\alpha)=\alpha^r.$$

The value of the numerical invariant μ can be used to give a criterion for the stability of a point, as in the following fundamental result.

Theorem 2.18 (Mumford's Numerical Criterion). *Let G be a reductive group acting on a projective variety X, let $\mathcal{L}\in\operatorname{Pic}^G(X)$ be an ample G-invariant line bundle, and let $x\in X$. Then*

$$x\in X^{\text{ss}}(\mathcal{L}) \iff \mu^{\mathcal{L}}(x,\ell)\ge 0 \text{ for all 1-parameter subgroups } \ell,$$
$$x\in X^{\text{s}}(\mathcal{L}) \iff \mu^{\mathcal{L}}(x,\ell)> 0 \text{ for all 1-parameter subgroups } \ell.$$

PROOF. See [88, Chapter 2, Theorem 2.1]. □

Example 2.19. Take

$$X=\mathbb{P}^2,\quad G=\mathrm{SL}_3,\quad \mathcal{L}=\mathcal{O}_{\mathbb{P}^2}(1).$$

Consider the 1-parameter subgroup

$$\ell\colon \mathbb{G}_m\to\mathrm{SL}_3,\quad \ell(\alpha)=\begin{pmatrix}\alpha&0&0\\0&\alpha^2&0\\0&0&\alpha^{-3}\end{pmatrix}.$$

The specialization of a point $x=[a,b,c]$ is

$$\ell(0)\cdot[a,b,c]=\lim_{\alpha\to0}[\alpha a,\alpha^2 b,\alpha^{-3}c]=\begin{cases}[0,0,1] & \text{if } c\ne 0,\\ [1,0,0] & \text{if } c=0 \text{ and } a\ne 0,\\ [0,1,0] & \text{if } a=c=0.\end{cases}$$

The stalk above $[0,0,1]$ is the "line" $\{(0,0,c) : c\in k\}$, and the action of $\ell(\alpha)$ on that line is via α^{-3}. Similarly the action on the stalk above $[1,0,0]$ is via α, and the action on the stalk above $[0,1,0]$ is via α^2. Hence (note that $\mu^{\mathcal{L}}$ is $-r$, not $+r$)

$$\mu^{\mathcal{O}(1)}([a,b,c],\ell)=\begin{cases}3 & \text{if } c\ne 0,\\ -1 & \text{if } c=0 \text{ and } a\ne0,\\ -2 & \text{if } a=c=0.\end{cases}$$

It might seem hard to apply Mumford's numerical criterion, because we need to compute $\mu^{\mathcal{L}}(x,\ell)$ for all 1-parameter subgroups ℓ. However, when X is projective space $\mathbb{P}^N$, it is always possible to diagonalize the action, because the linear action of a torus $\mathbb{G}_m^r$ on affine space can be diagonalized. For our 1-parameter subgroups, this leads to the following:

Theorem 2.20. *Let $G \subset \mathrm{SL}_{N+1}$, let $x \in \mathbb{P}^N$, let $\ell\colon \mathbb{G}_m \to G$ be a 1-parameter subgroup of G, and fix coordinates on $\mathbb{A}^{N+1}$ so that the action of $\mathbb{G}_m$ via ℓ is diagonalized, say*

$$\ell(\alpha) = \begin{pmatrix} \alpha^{r_0} & & & & \\ & \alpha^{r_1} & & & \\ & & \alpha^{r_2} & & \\ & & & \ddots & \\ & & & & \alpha^{r_N} \end{pmatrix}.$$

(Note that $r_0 + r_1 + \cdots + r_N = 0$.) Choose a point

$$\hat{x} = (\hat{x}_0, \hat{x}_1, \ldots, \hat{x}_N) \in \mathbb{A}^{N+1}$$

lying over x. Then relative to the ample sheaf $\mathcal{O}_{\mathbb{P}^N}(1)$,

$$\mu(x,\ell) = \max\{-r_i : i \text{ such that } \hat{x}_i \neq 0\}.$$

PROOF. See [88, Chapter 2, Proposition 2.3]. □

Example 2.21. Continuing with Example 2.19 in which SL_3 acts on $\mathbb{P}^2$, every diagonalized 1-parameter subgroup looks like

$$\ell_{ijk}\colon \mathbb{G}_m \to \mathrm{SL}_3, \quad \ell_{ijk}(\alpha) = \begin{pmatrix} \alpha^i & 0 & 0 \\ 0 & \alpha^j & 0 \\ 0 & 0 & \alpha^k \end{pmatrix}$$

for some $i,j,k \in \mathbb{Z}$ (not all zero) with $i+j+k=0$. In particular, one of i,j,k is nonpositive, so

$$\mu\bigl([a,b,c],\ell_{ijk}\bigr) > 0 \quad \text{provided } abc \neq 0.$$

On the other hand, if $abc = 0$, it's easy to find an ℓ_{ijk} that makes $\mu < 0$.

Thus a point $P \in \mathbb{P}^2$ is unstable if and only if, *after a change of coordinates*, one or more of its coordinates vanishes. But given any P, we can always find an $f \in \mathrm{SL}_3$ with (say) $f(P) = [0,b,c]$. Hence every point in $\mathbb{P}^2$ is unstable.

Example 2.22. We continue with the previous example, but we now take $G \subset \mathrm{SL}_3$ to be the subgroup consisting of diagonal matrices,

$$G = \left\{ \begin{pmatrix} u & 0 & 0 \\ 0 & v & 0 \\ 0 & 0 & w \end{pmatrix} \subset \mathrm{SL}_3 : uvw = 1 \right\}.$$

Then all 1-parameter subgroups of G are automatically diagonalized, so using the fact from Example 2.19 that

$$\mu\bigl([a,b,c],\ell_{ijk}\bigr) > 0 \iff abc \neq 0,$$

we find that for the action of G on $\mathbb{P}^2$,

$$(\mathbb{P}^2)^{\mathrm{ss}} = (\mathbb{P}^2)^{\mathrm{s}} = \bigl\{[a,b,c] \in \mathbb{P}^2 : abc \neq 0\bigr\}.$$

It turns out that the group G acts transitively on $(\mathbb{P}^2)^{\mathrm{s}}$, so (geometrically)

$$(\mathbb{P}^2)^{\mathrm{s}}/G = \bigl\{[1,1,1]\bigr\}$$

consists of a single point. To see that G acts transitively, we use the diagonal matrix

$$\begin{pmatrix} bu & 0 & 0 \\ 0 & au & 0 \\ 0 & 0 & (abu^2)^{-1} \end{pmatrix}$$

to send $[a, b, c]$ to $[abu, abu, c/abu^2]$. Over an algebraically closed field, we can take $u = (c/a^2b^2)^{1/3}$ to send $[a, b, c]$ to $[1, 1, 1]$.

2.3. Construction of moduli spaces M_d^n using GIT

Now it's time to apply the general GIT machinery to our situation. Fix $d \geq 2$ and $n \geq 1$. Recall from (1.3) that we have an embedding

$$\mathrm{Hom}_d^n \subset \mathbb{P}^N$$

for some large N, where we identify a map $\phi \in \mathrm{Hom}_d^n$ with a point in $\mathbb{P}^N$ by taking all of the coefficients of the monomials in the polynomials defining ϕ.

We let SL_{n+1} act on Hom_d^n via the usual conjugation,

$$\phi^f = f^{-1} \circ \phi \circ f \quad \text{for } \phi \in \mathrm{Hom}_d^n \text{ and } f \in \mathrm{SL}_{n+1}.$$

In terms of the embedding $\mathrm{Hom}_d^n \subset \mathbb{P}^N$, this gives an injective homomorphism

$$\mathrm{SL}_{n+1} \longrightarrow \mathrm{SL}_{N+1},$$

so we can apply GIT to the reductive group $G = \mathrm{SL}_{n+1} \subset \mathrm{SL}_{N+1}$ acting on the variety $X = \mathrm{Hom}_d^n \subset \mathbb{P}^N$. We are interested in knowing if categorical or geometric quotient varieties exist.

Example 2.23. The formula giving the action of SL_{n+1} on $\mathbb{P}^N$ is extremely complicated. We will explicitly describe the action of SL_2 on Rat_2. Let

$$f = \begin{pmatrix} \alpha & \beta \\ \gamma & \delta \end{pmatrix} \in \mathrm{SL}_2 \quad \text{and} \quad \phi = [a_0, a_1, a_2, b_0, b_1, b_2] \in \mathrm{Rat}_2 \subset \mathbb{P}^5,$$

so ϕ represents the rational map

$$\phi = \frac{F_a(x, y)}{F_b(x, y)} = \frac{a_0x^2 + a_1xy + a_2y^2}{b_0x^2 + b_1xy + b_2y^2} \in \mathrm{Rat}_2 \,.$$

In order to describe the action of f on ϕ, we compute

$$\begin{aligned} \phi^f &= f^{-1} \circ \phi \circ f \\ &= \frac{\delta F_a(\alpha x + \beta y, \gamma x + \delta y) - \beta F_b(\alpha x + \beta y, \gamma x + \delta y)}{-\gamma F_a(\alpha x + \beta y, \gamma x + \delta y) + \alpha F_b(\alpha x + \beta y, \gamma x + \delta y)} \\ &= \frac{A_0x^2 + A_1xy + A_2y^2}{B_0x^2 + B_1xy + B_2y^2}, \end{aligned}$$

where $A_0, \ldots, B_2$ are linear combinations of $a_0, \ldots, b_2$ with coefficients in the ring $k[\alpha, \beta, \gamma, \delta]$. For example,

$$A_0 = \delta\alpha^2 a_0 + \delta\alpha\gamma a_1 + \delta\gamma^2 a_2 - \beta\alpha^2 b_0 - \beta\alpha\gamma b_1 - \beta\gamma^2 b_2.$$

In this way the matrix $f \in \mathrm{SL}_2$ determines the matrix

$$g_f = \begin{pmatrix} \alpha^2\delta & \alpha\gamma\delta & \gamma^2\delta & -\alpha^2\beta & -\alpha\beta\gamma & -\beta\gamma^2 \\ 2\alpha\beta\delta & \alpha\delta^2+\beta\gamma\delta & 2\gamma\delta^2 & -2\alpha\beta^2 & -\alpha\beta\delta-\beta^2\gamma & -2\beta\gamma\delta \\ \beta^2\delta & \beta\delta^2 & \delta^3 & -\beta^3 & -\beta^2\delta & -\beta\delta^2 \\ -\alpha^2\gamma & -\alpha\gamma^2 & -\gamma^3 & \alpha^3 & \alpha^2\gamma & \alpha\gamma^2 \\ -2\alpha\beta\gamma & -\alpha\gamma\delta-\beta\gamma^2 & -2\gamma^2\delta & 2\alpha^2\beta & \alpha^2\delta+\alpha\beta\gamma & 2\alpha\gamma\delta \\ -\gamma\beta^2 & -\beta\gamma\delta & -\gamma\delta^2 & \alpha\beta^2 & \alpha\beta\delta & \alpha\delta^2 \end{pmatrix}$$

in SL_6, and the homomorphism

$$\mathrm{SL}_2 \longrightarrow \mathrm{SL}_6, \quad f \longmapsto g_f,$$

defines the action of SL_2 on $\mathbb{P}^5$ that we want to study. (Exercise: Check that $\det(g_f) = \det(f)^9$.) In this instance the entries of g_f are homogeneous polynomials of degree 3 in the entries $\alpha, \beta, \gamma, \delta$ of f, so even in this case of low dimension and low degree, the map $\mathrm{SL}_2 \to \mathrm{SL}_6$ and the action of SL_2 on $\mathrm{Rat}_2 \subset \mathbb{P}^6$ are quite complicated.

We begin by applying Mumford's criterion to the space of self-maps of $\mathbb{P}^1$. Let SL_2 act on $\mathbb{P}^{2d+1}$ via the conjugation action of SL_2 on $\mathrm{Hom}_d \subset \mathbb{P}^{2d+1}$, where the action of SL_2 on Hom_d is via the map $\mathrm{SL}_2 \to \mathrm{PGL}_2 = \mathrm{Aut}(\mathbb{P}^1)$. We write

$$(\mathrm{Hom}_d)^{\mathrm{s}} \quad \text{and} \quad (\mathrm{Hom}_d)^{\mathrm{ss}}$$

for the stable, respectively semistable, loci of this action, and we let

$$(\mathsf{M}_d)^{\mathrm{s}} = (\mathrm{Hom}_d)^{\mathrm{s}} / \mathrm{SL}_2 \quad \text{and} \quad (\mathsf{M}_d)^{\mathrm{ss}} = (\mathrm{Hom}_d)^{\mathrm{ss}} / \mathrm{SL}_2$$

be the corresponding geometric and categorical quotients.

Theorem 2.24 (**[109]**). *The set of morphisms* Hom_d *is invariant under the actionof* SL_2 *and is in the stable locus,*

$$\mathrm{Hom}_d \subset (\mathrm{Hom}_d)^{\mathrm{s}}.$$

Hence there is a geometric quotient space

$$\mathsf{M}_d = \mathrm{Hom}_d / \mathrm{SL}_2$$

with structure sheaf

$$\mathcal{O}_{\mathsf{M}_d} = \left(\mathcal{O}_{\mathrm{Hom}_d}\right)^{\mathrm{SL}_2}.$$

There are natural inclusions

$$\mathsf{M}_d \subset (\mathsf{M}_d)^{\mathrm{s}} \subset (\mathsf{M}_d)^{\mathrm{ss}}.$$

Further

$$(\mathsf{M}_d)^{\mathrm{s}} = (\mathsf{M}_d)^{\mathrm{ss}} \iff d \textit{ is even.}$$

Remark 2.25. For $d \geq 2$ we have

$$\dim \mathrm{Hom}_d = 2d+1, \quad \dim \mathrm{SL}_2 = 3, \quad \dim \mathsf{M}_d = 2d-2.$$

PROOF. First we observe that conjugation simply moves the zeros and poles of a rational map ϕ, so

$$\deg \phi^f = \deg \phi.$$

Hence Hom_d is invariant under SL_2-conjugation. That's the easy part of the theorem.

We want to determine when a map $\phi \in \mathrm{Hom}_d$ corresponds to a stable, semi-stable or unstable point in $\mathbb{P}^{2d+1}$. The map ϕ is associated to a point in $\mathbb{P}^{2d+1}$ via the usual identification

$$\phi = \bigl[F_a(x,y), F_b(x,y)\bigr] = \frac{a_0x^d + a_1x^{d-1}y + a_2x^{d-2}y^2 + \cdots + a_dy^d}{b_0x^d + b_1x^{d-1}y + b_2x^{d-2}y^2 + \cdots + b_dy^d}$$
$$\updownarrow$$
$$[a_0, \dots, a_d, b_0, \dots, b_d] \in \mathbb{P}^{2d+1}.$$

Let

$$\ell\colon \mathbb{G}_m \longrightarrow \mathrm{SL}_2$$

be a 1-parameter family. Conjugating by an appropriate element of SL_2, we can diagonalize ℓ, and if desired, switch the two coordinates, so ℓ will have the form

$$\ell_k\colon \mathbb{G}_m \longrightarrow \mathrm{SL}_2, \quad \ell_k(\alpha) = \begin{pmatrix} \alpha^k & 0 \\ 0 & \alpha^{-k} \end{pmatrix} \qquad \text{for some } k \in \mathbb{Z},\ k \ge 1.$$

This also diagonalizes the action of SL_2 on $\mathbb{A}^{2d+2}$. More precisely, writing

$$\begin{aligned} \ell_k(\alpha)\cdot\phi &= \ell_k(\alpha)^{-1} \circ [F_a, F_b] \circ \ell_k(\alpha) \\ &= \bigl[\alpha^{-k}F_a(\alpha^k x, \alpha^{-k}y), \alpha^k F_b(\alpha^k x, \alpha^{-k}y)\bigr], \end{aligned}$$

we have

$$\begin{aligned} a_ix^iy^{d-i} &\longmapsto \alpha^{-k}a_i(\alpha^k x)^i(\alpha^{-k}y)^{d-i} = \alpha^{k(2i-d-1)}a_ix^iy^{d-i}, \\ b_ix^iy^{d-i} &\longmapsto \alpha^{k}b_i(\alpha^k x)^i(\alpha^{-k}y)^{d-i} = \alpha^{k(2i-d+1)}b_ix^iy^{d-i}. \end{aligned}$$

Thus

$$\ell_k(\alpha)\cdot\phi = \ell_k(\alpha)\cdot[a_0, \dots, a_d, b_0, \dots, b_d]$$

is given explicitly on the coordinates $[a_0, \dots, b_d]$ of $\mathbb{P}^{2d+1}$ by the formulas

$$a_i \longmapsto \alpha^{k(2i-d-1)}a_i, \quad b_i \longmapsto \alpha^{k(2i-d+1)}b_i.$$

The numerical criterion (Theorem 2.18) says:

$$\begin{aligned} \phi = (a_0, \dots, b_d) \text{ is unstable} &\iff \mu^{\mathcal{O}(1)}(\phi, \ell_k) < 0 \quad \text{for some } k \ge 1. \\ \phi = (a_0, \dots, b_d) \text{ is not stable} &\iff \mu^{\mathcal{O}(1)}(\phi, \ell_k) \le 0 \quad \text{for some } k \ge 1. \end{aligned}$$

Further, since $\ell_k(\alpha) = \ell_1(\alpha)^k$, we have

$$\mu^{\mathcal{O}(1)}(\phi, \ell_k) = k\mu^{\mathcal{O}(1)}(\phi, \ell_1),$$

so it's enough to look at ℓ_1.

Recall the characterization

$$\mu^{\mathcal{O}(1)}(x, \ell) = \max\{-r_i : i \text{ such that } \hat{x}_i \ne 0\}.$$

In our situation, this becomes

$$\mu^{\mathcal{O}(1)}(\phi, \ell_1) = \max\bigl(\{d+1-2i : a_i \ne 0\} \cup \{d-1-2i : b_i \ne 0\}\bigr).$$

Hence ϕ is unstable if and only if

$$a_i \ne 0 \implies d+1-2i < 0 \quad \text{and} \quad b_i \ne 0 \implies d-1-2i < 0.$$

Equivalently,

$$\phi \text{ is unstable} \iff a_i = 0 \text{ for all } i \le \frac{d+1}{2} \text{ and } b_i = 0 \text{ for all } i \le \frac{d-1}{2}. \tag{2.2}$$

The condition for not stable, i.e., the union of semistable or unstable, simply requires replacing the inequalities by strict inequalities. Thus

$$\text{(2.3)}\quad \phi \text{ is not stable} \iff a_i = 0 \text{ for all } i < \frac{d+1}{2} \text{ and } b_i = 0 \text{ for all } i < \frac{d-1}{2}.$$

However, it is important to observe that conditions (2.2) or (2.3) are after a change of coordinates used to diagonalize the action of the 1-parameter subgroup. So they really say that ϕ is unstable if there exists some change of coordinates ϕ^f so that the coefficients of the numerator and denominator of ϕ^f satisfy (2.2), and similarly for semistable using (2.3).

These characterizations immediately give us the desired results. Thus

$$\phi \text{ not stable} \implies a_0 = b_0 = 0 \text{ for some } \phi^f \implies \phi \notin \mathrm{Hom}_d,$$

which proves that $\mathrm{Hom}_d \subset (\mathrm{Hom}_d)^{\mathrm{s}}$. It then follows from GIT (Theorem 2.15) that there is a geometric quotient space

$$\mathsf{M}_d = \mathrm{Hom}_d \,/\, \mathrm{SL}_2 \quad \text{with structure sheaf } \mathcal{O}_{\mathsf{M}_d} = (\mathcal{O}_{\mathrm{Hom}_d})^{\mathrm{SL}_2}.$$

Finally, if d is even, then the condition

$$i \le \frac{d \pm 1}{2} \quad \text{is the same as} \quad i < \frac{d \pm 1}{2},$$

so the unstable and not stable conditions coincide. On the other hand, if d is odd, then the two conditions do not coincide, and it is easy to write down maps that are semistable, but not stable. □

The construction of dynamical moduli spaces can be generalized to self-maps of higher dimensional projective spaces.

Theorem 2.26 (Petsche–Szpiro–Tepper [95], Levy [66]). *Let* SL_{n+1} *act on* $\mathbb{P}^N$ *via the conjugation action of* SL_{n+1} *on* $\mathrm{Hom}_d^n \subset \mathbb{P}^N$*. Then* Hom_d^n *is in the stable locus, and hence there is a geometric quotient variety*

$$\mathsf{M}_d^n = \mathrm{Hom}_d^n \,/\, \mathrm{SL}_{n+1}$$

with the usual properties.

Proof Sketch. As before, SL_{n+1} conjugation simply moves around any common zeros of the coordinate functions of ϕ, so ϕ is a morphism if and only if ϕ^f is a morphism. Hence Hom_d^n is SL_{n+1}-invariant.

Now let $\ell\colon \mathbb{G}_m \to \mathrm{SL}_{n+1}$ be a 1-parameter subgroup, and diagonalize the action on the coordinates of $\mathbb{P}^n$. Thus there are integers $r_0, \ldots, r_n$ with

$$r_0 + r_1 + \cdots + r_n = 0$$

such that

$$\ell(\alpha) \cdot x_i = \alpha^{r_i} x_i \quad \text{for } 0 \le i \le n.$$

From this we see that the coefficients of $\ell(\alpha) \cdot \phi$ are equal to the coefficients of ϕ multiplied by some power of α (with different coefficients multiplied by different powers). Thus the action is also diagonalized on $\mathbb{P}^N \supset \mathrm{Hom}_d^n$.

More precisely, if we write

$$\phi = [\phi_0, \ldots, \phi_n], \quad \phi_u = \sum a_{ue_{u0}\ldots e_{un}} x_0^{e_{u0}} x_1^{e_{u1}} \cdots x_n^{e_{un}},$$

then the action of $\ell(\alpha)$ on the $a_{ue_{u0}\ldots e_{un}}$ coordinate of $\mathbb{P}^N$ is given by

$$\ell(\alpha) \cdot a_{ue_{u0}\ldots e_{un}} = \alpha^{-r_u} \alpha^{r_0 e_{u0} + r_1 e_{u1} + \cdots + r_n e_{un}} a_{ue_{u0}\ldots e_{un}}.$$

TABLE 2.1

	$d=1$	$d=2$	$d=3$	$d=4$	$d=5$	$d=6$
$n=1$	2	4	6	8	10	12
$n=2$	9	21	36	54	75	99
$n=3$	24	64	124	208	320	464
$n=4$	50	150	325	605	1025	1625

Now a careful analysis gives a finite collection of linear inequalities on the subscripts e_{ui} such that

$$\begin{aligned} &\phi \text{ is unstable} \\ &\qquad \iff a_{ue_{u0}\cdots e_{un}} = 0 \text{ for all } e_{u0},\ldots,e_{un} \text{ satisfying some linear inequalities.} \end{aligned}$$

In particular, if ϕ is unstable, then one finds that $\phi_0,\ldots,\phi_n$ have a common zero, so $\phi \notin \mathrm{Hom}_d^n$. See [66] for details. □

Remark 2.27. We have

$$\begin{aligned} \dim \mathrm{Hom}_d^n &= \binom{n+d}{d}(n+1) - 1, \\ \dim \mathrm{SL}_{n+1} &= (n+1)^2 - 1, \\ \dim \mathsf{M}_d^n &= \binom{n+d}{d}(n+1) - (n+1)^2. \end{aligned}$$

Table 2.1 gives the dimension of M_d^n for small values of n and d.

Remark 2.28. The semistable and stable quotients each have certain important features which come from general results in geometric invariant theory. In particular, for an algebraically closed field k,

$$(\mathsf{M}_d^n)^{\mathrm{s}}(k) = \frac{(\mathrm{Hom}_d^n)^{\mathrm{s}}(k)}{\mathrm{SL}_{n+1}(k)} \quad \text{and} \quad (\mathsf{M}_d^n)^{\mathrm{ss}}(k) \text{ is proper (complete) over } k.$$

For $n=1$ and $d \equiv 0 \pmod 2$, Theorem 2.24 says that the stable and semistable loci coincide, so $(\mathsf{M}_d^1)^{\mathrm{s}} = (\mathsf{M}_d^1)^{\mathrm{ss}}$ has both properties. This does not happen for any larger values of n; see [66].

2.4. Multipliers and maps on M_d

The GIT construction gives us the existence of M_d^n, but GIT tells us little about the structure of M_d^n as a variety. In this section we discuss the case $n=1$. We will explicitly construct functions on Hom_d that are PGL_2-conjugation invariant, and then we will use these functions to map M_d to affine space.

Definition 2.29. Let $\phi(z) \in K(z)$, and let $\alpha \in \mathrm{Per}(\phi)$ be a point of exact period n. The *multiplier of ϕ at α* is

$$\lambda_\phi(\alpha) = (\phi^n)'(\alpha) = \prod_{i=0}^{n-1} \phi'(\phi^i(\alpha)).$$

This definition of $\lambda_\phi(\alpha)$ makes sense provided $\infty \notin \mathcal{O}_\phi(\alpha)$. For the general case, one can use the following lemma to move points away from ∞, or see Remark 2.32 for an intrinsic definition of $\lambda_\phi(\alpha)$.

Lemma 2.30. *Let $\alpha \in \mathrm{Per}_n(\phi)$ be a point of exact period n and let $f \in \mathrm{PGL}_2$. Then*

$$\mathrm{Per}(\phi) = f^{-1}\bigl(\mathrm{Per}(\phi^f)\bigr)$$

and

$$\lambda_\phi(\alpha) = \lambda_{\phi^f}\bigl(f^{-1}(\alpha)\bigr).$$

PROOF. The first statement is obvious, and the second is an exercise using the chain rule. □

The set of multipliers, taken with appropriate multiplicity, is called the *multiplier spectrum of ϕ* and is denoted by

$$\Lambda_\phi^n = \bigl\{\lambda_\phi(\alpha) : \alpha \in \mathrm{Per}_n(\phi)\bigr\}.$$

Lemma 2.30 says that the set Λ_ϕ^n (with multiplicities) depends only on the PGL_2-equivalence class of ϕ. If ϕ has degree d, then counted with appropriate multiplicities, the set Λ_ϕ^n contains $d^n + 1$ points. Thus if K is an algebraically closed field, then there is a well-defined map

$$\mathsf{M}_d(K) \longrightarrow \begin{Bmatrix} \text{multisets of } K \text{ with} \\ d^n+1 \text{ elements} \end{Bmatrix}, \qquad \langle\phi\rangle \longmapsto \Lambda_\phi^n.$$

Proposition 2.31. *Let $\phi \in \mathrm{Hom}_d$. Then ϕ has d^n+1* distinct *n-periodic points if and only if $1 \notin \Lambda_\phi^n$.*

PROOF. Replacing ϕ with ϕ^n, it suffices to look at fixed points, and changing coordinates, we may assume that ∞ is not a fixed point. Then the fixed points of ϕ are the zeros of the rational function $\phi(z) - z$. We claim that $\alpha \in \mathrm{Fix}(\phi)$ is a multiple root of $\phi(z) - z$ if and only if $\phi'(\alpha) = 1$. To see this, consider the Taylor expansion of $\phi(z)$ around α,

$$\phi(z) = \phi(\alpha) + \phi'(\alpha)(z - \alpha) + O\bigl((z-\alpha)^2\bigr).$$

Using the assumption that α is a fixed point, i.e., $\phi(\alpha) = \alpha$, this formula becomes

$$\phi(z) - z = (\phi'(\alpha) - 1)(z - \alpha) + O\bigl((z-\alpha)^2\bigr),$$

which gives the desired result. □

Remark 2.32. If $\infty \in \mathcal{O}_\phi(\alpha)$, we can use Lemma 2.30 to change coordinates before computing $\lambda_\phi(\alpha)$. But a better, more intrinsic, way to think of the multiplier is as a map on the tangent space. Thus if $P \in \mathrm{Per}_n(\phi)$, then ϕ^n induces a map on the tangent space

$$(\phi_P^n)_* \colon \mathcal{T}_P(\mathbb{P}^1) \longrightarrow \mathcal{T}_P(\mathbb{P}^1).$$

The tangent space is 1-dimensional, so

$$(\phi_P^n)_* \in \mathrm{Aut}\bigl(\mathcal{T}_P(\mathbb{P}^1)\bigr) = \mathrm{Aut}(\mathbb{A}^1) = \mathbb{G}_m.$$

Thus $(\phi_P^n)_*$ is a number, and this number is the multiplier $\lambda_\phi(\alpha)$.

A rational map $\phi \in \mathbb{Q}(a_0, \ldots, a_d, b_0, \ldots, b_d)(z)$ is defined over the field generated by its coefficients, but in general, the multipliers of the n-periodic points are only defined over an algebraic extension of

$$\mathbb{Q}(a_0, \ldots, a_d, b_0, \ldots, b_d).$$

In order to get elements of $\mathbb{Q}(a_0,\ldots,a_d,b_0,\ldots,b_d)$, we take symmetric functions of the multipliers over all n-periodic points. Thus we define

$$\sigma_{i,n}(\phi) \in \mathbb{Q}(a_0,\ldots,a_d,b_0,\ldots,b_d)$$

by the formula

$$\prod_{P\in\mathrm{Per}_n(\phi)} (T+\lambda_\phi(P)) = \sum_{i=0}^{d^n+1} \sigma_{i,n}(\phi)T^{d^n+1-i}.$$

Then

$$\sigma_{i,n}(\phi^f) = \sigma_{i,n}(\phi) \quad \text{for all } f \in \mathrm{SL}_2,$$

so

$$\sigma_{i,n} \in \mathbb{Q}(a_0,\ldots,a_d,b_0,\ldots,b_d)^{\mathrm{SL}_2} = \mathbb{Q}(\mathsf{M}_d).$$

Further, the function

$$\sigma_{i,n} \colon \mathsf{M}_d \longrightarrow \mathbb{A}^1$$

is well-defined at every point of M_d, i.e., it is a regular function on M_d, so it is in the affine coordinate ring of M_d,

$$\sigma_{i,n} \in \mathbb{Q}[\mathsf{M}_d].$$

Definition 2.33. The functions $\sigma_{i,n}$ are called *Milnor multiplier functions* on the moduli space M_d.

Remark 2.34. One can get finer invariants using only points of "formal period n." These are (roughly) the roots of the dynatomic polynomial

$$\prod_{e|n} (\phi^e(z) - z)^{\mu(n/e)},$$

which is a dynamical analogue of the nth cyclotomic polynomial. (Here μ is the Möbius function.)See Section 4.1.

Example 2.35. Let $\phi(z) = z^d$, so $\mathrm{Per}_1(\phi) = \boldsymbol{\mu}_{d-1} \cup \{0,\infty\}$. Then

$$\lambda_\phi(\zeta) = d\zeta^{d-1} = d \quad \text{and} \quad \lambda_\phi(0) = \lambda_\phi(\infty) = 0,$$

which gives

$$\prod_{P\in\mathrm{Per}_1(\phi)} (T+\lambda_\phi(P)) = (T+d)^{d-1}T = \sum_{i=0}^{d-1} \binom{d}{i} d^i T^{d-i+2}.$$

From this one can read off the values of $\sigma_{i,n}(z^d)$.

The multipliers of Lattès maps are interesting, because they give examples of inequivalent rational maps whose multiplier sets $\Lambda_{\phi,n}$ are identical. Lattès maps are discussed further in Theorem 2.38, Remark 2.39, and Section 3.4.

Proposition 2.36. *Let E be an elliptic curve, and let $\phi = \phi_{E,m}$ be the flexible Lattès map associated to the multiplication-by-m map on E. Thus ϕ fits into the commutative square*

$$\begin{array}{ccc} E & \xrightarrow{\;m\;} & E \\ \downarrow{\scriptstyle x} & & \downarrow{\scriptstyle x} \\ \mathbb{P}^1 & \xrightarrow{\;\phi\;} & \mathbb{P}^1. \end{array}$$

(a) *The set of n periodic points of $\phi = \phi_{E,m}$ is*

$$\operatorname{Per}_n(\phi) = x\bigl(E[m^n - 1]\bigr) \cup x\bigl(E[m^n + 1]\bigr).$$

(b) *Let $P \in \operatorname{Per}_n(\phi)$ be a point of exact period n. Then*

$$\lambda_\phi(P) = \begin{cases} m^n & \text{if } P \in x\bigl(E[m^n-1]\bigr) \text{ and } P \notin x\bigl(E[2]\bigr), \\ -m^n & \text{if } P \in x\bigl(E[m^n+1]\bigr) \text{ and } P \notin x\bigl(E[2]\bigr), \\ m^{2n} & \text{if } P \in x\bigl(E[m^n+1]\bigr) \cap x\bigl(E[2]\bigr). \end{cases}$$

(c) *The value of the Milnor function $\sigma_{i,n}(\phi_{E,m})$ depends only on i, n, and m. It is independent of E.*

PROOF. Apply the relation $x \circ [m] = \phi \circ x$ to the invariant differential $\omega = dx/2y$ and use the fact that $[m]^*\omega = m\omega$. See [110, Proposition 6.52] for details. □

Example 2.37. Let $\phi(z) = (a_0 z^2 + a_1 z + a_2)/(b_0 z^2 + b_1 z + b_2)$. Then

$$\sigma_{1,1}(\phi) = \frac{\text{huge mess in } a_i\text{'s and } b_i\text{'s}}{\text{resultant}}.$$

Here is the actual formula:

$$\sigma_{1,1}(\phi) = \frac{\begin{array}{r} a_1^3 b_0 - 4a_0a_1a_2b_0 - 6a_2^2b_0^2 - a_0a_1^2b_1 + 4a_0^2a_2b_1 + 4a_1a_2b_0b_1 - 2a_0a_2b_1^2 + a_2b_1^3 \\ - 2a_1^2b_0b_2 + 4a_0a_2b_0b_2 - 4a_2b_0b_1b_2 - a_1b_1^2b_2 + 2a_0^2b_2^2 + 4a_1b_0b_2^2 \end{array}}{a_2^2b_0^2 - a_1a_2b_0b_1 + a_0a_2b_1^2 + a_1^2b_0b_2 - 2a_0a_2b_0b_2 - a_0a_1b_1b_2 + a_0^2b_2^2}.$$

As noted earlier, the Milnor functions $\sigma_{i,n}$ are in the affine coordinate ring

$$\mathbb{Q}[\mathsf{M}_d] = \mathbb{Q}[\operatorname{Hom}_d]^{\operatorname{SL}_2},$$

and in fact if we construct M_d over $\mathbb{Z}$, then they are in $\mathbb{Z}[\mathsf{M}_d]$. We can thus use them to map M_d into affine space. A beautiful theorem of McMullen says that this map is (almost) finite-to-one.

Theorem 2.38 (McMullen [73]). *Let*

$$\boldsymbol{s}_N : \mathsf{M}_d \longrightarrow \mathbb{A}^L \tag{2.4}$$

be the morphism defined using all of the Milnor multiplier functions $\sigma_{i,n}$ with $1 \le n \le N$ and $1 \le i \le d^n + 1$. If N is sufficiently large, then the map

$$\boldsymbol{s}_N : \mathsf{M}_d(\mathbb{C}) \longrightarrow \mathbb{C}^L$$

is finite-to-one except when $n = m^2$ is a perfect square, in which case it is finite-to-one except that Lattès maps associated to multiplication-by-m maps on elliptic curves are all sent to the same point.

Remark 2.39. We will see in Section 2.5 that $\boldsymbol{s}_1 : \mathsf{M}_2 \to \mathbb{A}^3$ identifies M_2 with a hyperplane in $\mathbb{A}^3$. On the other hand, using rigid Lattès maps and classical estimates for the class number of $\mathbb{Q}(\sqrt{-d}\,)$, it is not hard to show that for all $\epsilon > 0$, all sufficiently large d, and all N, there exist points $\xi \in \mathsf{M}_d(\mathbb{C})$ such that

$$\#\, \boldsymbol{s}_N^{-1}\bigl(\boldsymbol{s}_N(\xi)\bigr) \ge C_{N,\epsilon} d^{1/2-\epsilon}.$$

So for large values of d, the map $\boldsymbol{s}_N$ is far from being one-to-one. See [110, Theorem 6.62] for details.

Remark 2.40. We briefly describe one way to construct Milnor multiplier functions in dimension greater than one. Let $\phi \in \mathrm{Hom}_d^N$ and $P \in \mathrm{Per}_n(\phi)$, and consider the map on the tangent space (cf. Remark 2.32)

$$(\phi_P^n)_* \colon \mathcal{T}_P(\mathbb{P}^N) \longrightarrow \mathcal{T}_P(\mathbb{P}^N),$$
$$(\phi_P^n)_* \in \mathrm{Aut}\bigl(\mathcal{T}_P(\mathbb{P}^N)\bigr) = \mathrm{Aut}(\mathbb{A}^N) \cong \mathrm{GL}_N\,.$$

The identification of the linear transformation $(\phi_P^n)_*$ with a matrix in GL_N requires choosing a basis for the tangent space $\mathcal{T}_P(\mathbb{P}^N)$, but the characteristic polynomial of $(\phi_P^n)_*$ is independent of this choice. So we can define a map

$$\lambda_\phi \colon \mathrm{Per}_n(\phi) \longrightarrow \mathbb{A}^N, \quad P \longmapsto \begin{pmatrix}\text{coefficients of the characteristic polynomial of}\\ \text{the linear transformation } (\phi_P^n)_*\end{pmatrix}.$$

Note that the coordinates of $\lambda_\phi(P)$, i.e., the coefficients of the characteristic polynomial of $(\phi_P^n)_*$, are defined over some finite extension of the field containing the coefficients of ϕ. In order to obtain values that are defined over the same field as ϕ, we proceed as in the one-dimensional case. Thus for each $1 \le i \le N$ we take the elementary symmetric functions of the ith coordinates of $\lambda_\phi(P)$ for all $P \in \mathrm{Per}_n(\phi)$. Each of these quantities depends rationally on the coefficients of ϕ and is SL_{N+1}-invariant. They thus define regular functions on M_d^N.

Remark 2.41. It is natural to ask whether McMullen's theorem (Theorem 2.38) is true in characteristic p? An obvious class of counter-examples are maps of the form $\phi(z) = \psi(z^p)$, since every multiplier of such a map vanishes. So we certainly want to restrict to separable maps. However, even with this restriction, the following example of Alon Levy shows that the Milnor map (2.4) is not even close to being finite-to-one in characteristic p.

Proposition 2.42 (Levy, personal communication). *Let K be a field of characteristic p. Let $F(z), G(z) \in K[z]$ be polynomials with no common roots, let $A \in K^*$, and let*

$$\phi(z) = Az + \frac{F(z^p)}{G(z^p)}.$$

Then

$$\phi'(\alpha) = A \quad \text{for all } \alpha \in \mathrm{Per}(\phi) \smallsetminus \{\infty\}. \tag{2.5}$$

Further, $\infty \in \mathrm{Fix}(\phi)$, and

$$\phi'(\infty) = \begin{cases} A^{-1} & \text{if } \deg F \le \deg G, \\ 0 & \text{if } \deg F > \deg G. \end{cases} \tag{2.6}$$

Hence the value $\boldsymbol{s}_N(\phi)$ of the Milnor map (2.4) depends only on A, and is independent of the choice of the polynomials F and G, other than $\lambda_\phi(\infty)$ depending on the relative magnitude of the degrees of F and G.

Proof. It is clear that $\phi(\infty) = \infty$. Hence the points in $\phi^{-1}(\infty)$, which are the roots of $G(z^p)$, are strictly preperiodic. Thus in order to verify (2.5), it suffices to compute $\phi'(z)$ using elementary differentiation rules and then substitute $z = \alpha$. Since K has characteristic p, we find that

$$\phi'(z) = A + \frac{\bigl(G(z^p)F'(z^p) - F(z^p)G'(z^p)\bigr)pz^{p-1}}{G(z^p)^2} = A.$$

Next, to compute $\phi'(\infty)$, we conjugate by $f(z) = z^{-1}$ and compute (note $f = f^{-1}$)

$$(\phi^f)'(z) = f'\Big(\phi\big(f(z)\big)\Big) \cdot \phi'\big(f(z)\big) \cdot f'(z) = \frac{A}{\phi(z^{-1})^2 z^2} = \frac{A}{\big(A + zF(z^{-p})/G(z^{-p})\big)^2}.$$

Hence

$$\phi'(\infty) = (\phi^f)'(0) = \begin{cases} A^{-1} & \text{if } \deg F \le \deg G, \\ 0 & \text{if } \deg F > \deg G. \end{cases}$$

Finally, we compute the multiplier of a point $\alpha \in \operatorname{Per}(\phi)$ of exact period n. If $\alpha = \infty$, then $\lambda_\phi(\infty) = A^{-1}$ or 0 as described by (2.6). If $\alpha \ne \infty$, then

$$\lambda_\phi(\alpha) = (\phi^n)'(\alpha) = \prod_{i=0}^{n-1} \phi'\big(\phi^i(\alpha)\big) = A^n. \qquad \square$$

Here are a few arithmetic questions centered around McMullen's theorem and the Milnor map.

Question 2.43. (a) Is it true, as suggested by Levy, that McMullen's theorem (Theorem 2.38) is true in characteristic p for families of maps having no wild ramification, i.e., no points where the ramification index is divisible by p?

(b) McMullen defines a stable family of rational maps $\phi_t(z)$ to be a family in which there is a bound N such that the period of every attracting fixed point of $\phi_t(z)$ is bounded by N, independent of t. A key step in McMullen's proof of Theorem 2.38 is to show that over $\mathbb{C}$, every stable family is either Lattès or PCF. What is the appropriate p-adic analogue of this statement, keeping in mind that p-adic rational maps may have infinitely many p-adically attracting periodic points?

(c) Is the analog of McMullen's theorem true for M_d^N with $N \ge 2$? (See Remark 2.40 for one way to define higher dimensional Milnor multiplier functions.)

(d) Is there a nonempty Zariski open subset $U \subset \mathsf{M}_d^1$ and an integer N such that $\boldsymbol{s}_N : U(\mathbb{C}) \to \mathbb{C}^L$ is one-to-one? (This question was posed by Bjorn Poonen at the workshop.)

Remark 2.44. We briefly describe another way to construct a conjugation invariant function on Hom_d, and thus a function on the moduli space M_d. The *Schwarzian derivative* of a rational map $\phi(z) \in K(z)$ is the function $\mathcal{S}\phi$ defined by the formula

$$(\mathcal{S}\phi)(z) = \frac{\phi'''(z)}{\phi'(z)} - \frac{3}{2}\left(\frac{\phi''(z)}{\phi'(z)}\right)^2.$$

Although not immediately apparent from the definition, one can show that the Schwarzian derivative induces a morphism

$$\operatorname{Hom}_d \times \mathbb{P}^1 \longrightarrow \mathbb{P}^1, \quad (\phi, P) \longmapsto (\mathcal{S}\phi)(P).$$

The Schwarzian derivative measures the difference between ϕ and the best approximation to ϕ by linear fractional transformations. In particular, the Schwarzian derivative of a linear fractional transformation $f \in \mathrm{PGL}_2$ satisfies $\mathcal{S}f = 0$, and the Schwarzian derivative of a PGL_2-conjugate is given by the chain rule

$$(\mathcal{S}\phi^f)(z) = f'(z)^2 \cdot (\mathcal{S}\phi)\big(f(z)\big).$$

It follows that the quadratic differential form

$$\omega_\phi(z) = (\mathcal{S}\phi)(z)(dz)^2$$

is PGL_2-invariant, i.e., $\omega_{\phi^f}(f^{-1}z) = \omega_\phi(z)$, so there is a well-defined map

$$\mathsf{M}_d \longrightarrow \begin{pmatrix}\text{holomorphic quadratic} \\ \text{differential forms on } \mathbb{P}^1\end{pmatrix}, \qquad \phi \longrightarrow \omega_\phi.$$

2.5. M_2 is isomorphic to $\mathbb{A}^2$

In this section we study rational maps of degree two on $\mathbb{P}^1$, i.e., the spaces Hom_2 and M_2. To ease notation, we write

$$\sigma_1 = \sigma_{1,1}, \qquad \sigma_2 = \sigma_{2,1}, \qquad \sigma_3 = \sigma_{3,1}$$

for the multiplier functions associated to the fixed points.

Theorem 2.45 (Milnor [78] over $\mathbb{C}$, Silverman [109] over $\mathbb{Z}$).

(a) $\sigma_1 = \sigma_3 + 2$.

(b) *The map*

$$s = (\sigma_1, \sigma_2) \colon \mathsf{M}_2 \longrightarrow \mathbb{A}^2$$

is an isomorphism.

PROOF. (a) Let $\phi \in \mathrm{Hom}_2(\mathbb{C})$ have multipliers $\lambda_1, \lambda_2, \lambda_3$, and assume that no multiplier is equal to 1. We use the well-known formula

$$\frac{1}{1-\lambda_1} + \frac{1}{1-\lambda_2} + \frac{1}{1-\lambda_3} = 1,$$

which is proven by applying Cauchy's residue theorem to the function

$$\frac{1}{\phi(z) - z}.$$

A bit of algebra yields

$$\lambda_1\lambda_2\lambda_3 - (\lambda_1 + \lambda_2 + \lambda_3) + 2 = 0, \qquad \text{i.e., } \sigma_3 - \sigma_1 + 2 = 0.$$

The set of ϕ with no multiplier equal to 1 is dense in Hom_2, so the relation $\sigma_3 - \sigma_1 + 2 = 0$ holds for all ϕ.

(b) We'll use the following.

Lemma 2.46 (Normal Forms Lemma). *Let*

$$\phi \in \mathrm{Hom}_2(\mathbb{C}) \quad \textit{and} \quad \Lambda_1(\phi) = \{\lambda_1, \lambda_2, \lambda_3\}.$$

(i) *If $\lambda_1\lambda_2 \ne 1$, then there is an $f \in \mathrm{SL}_2(\mathbb{C})$ such that*

$$\phi^f(z) = \frac{z^2 + \lambda_1 z}{\lambda_2 z + 1}.$$

The resultant of this normal form is

$$\mathrm{Resultant}(\phi^f) = 1 - \lambda_1\lambda_2.$$

(ii) *If $\lambda_1\lambda_2 = 1$, then $\lambda_1 = \lambda_2 = 1$ and there is an $f \in \mathrm{SL}_2(\mathbb{C})$ such that*

$$\phi^f(z) = z + \sqrt{1-\lambda_3} + \frac{1}{z}.$$

PROOF. Let $\alpha \in \mathrm{Fix}(\phi)$. Then

$$\begin{aligned}\phi(z) &= \phi(\alpha) + \phi'(\alpha)(z-\alpha) + O\big((z-\alpha)^2\big)\\ &= \alpha + \lambda_\phi(\alpha)(z-\alpha) + O\big(z-\alpha)^2\big),\end{aligned}$$

so

$$\phi(z) - z = \big(\lambda_\phi(\alpha) - 1\big)(z-\alpha) + O(z-\alpha)^2.$$

Thus the multiplicity of the fixed point α satisfies

$$\mathrm{Mult}(\phi,\alpha) = 1 \iff \lambda_\phi(\alpha) \neq 1. \tag{2.7}$$

We next apply the formal identity

$$(X-1)^2 - (XY-1)(XZ-1) = X(X+Y+Z-2-XYZ) \tag{2.8}$$

with $\{X,Y,Z\} = \{\lambda_1,\lambda_2,\lambda_3\}$. Using the relation

$$\lambda_1+\lambda_2+\lambda_3 = \sigma_1 = \sigma_3 + 2 = \lambda_1\lambda_2\lambda_3 + 2,$$

the right-hand side of (2.8) vanishes, which give us three useful formulas,

$$(\lambda_1-1)^2 = (\lambda_1\lambda_2-1)(\lambda_1\lambda_3-1), \tag{2.9}$$

$$(\lambda_2-1)^2 = (\lambda_2\lambda_1-1)(\lambda_2\lambda_3-1), \tag{2.10}$$

$$(\lambda_3-1)^2 = (\lambda_3\lambda_1-1)(\lambda_3\lambda_2-1). \tag{2.11}$$

In particular,

$$\lambda_1\lambda_2 = 1 \iff \lambda_1 = \lambda_2 = 1.$$

(i) Let $\mathrm{Fix}(\phi) = \{\alpha_1,\alpha_2,\alpha_3\}$. We are assuming that $\lambda_1\lambda_2 \neq 1$, and we have

$$\begin{aligned}\lambda_1\lambda_2 \neq 1 &\iff \lambda_1 \neq 1 \text{ and } \lambda_2 \neq 1 && \text{from (2.9) and (2.10)},\\ &\iff \mathrm{Mult}(\phi,\alpha_1) = \mathrm{Mult}(\phi,\alpha_2) = 1 && \text{from (2.7)}.\end{aligned}$$

In particular, $\alpha_1 \neq \alpha_2$, so we can find an $f \in \mathrm{SL}_2(\mathbb{C})$ that moves α_1 to 0 and α_2 to ∞. Then ϕ^f has the form

$$\phi^f(z) = \frac{a_0z^2 + a_1z}{b_1z + b_2} \quad \text{with } a_0b_2 \neq 0.$$

We dehomogenize by setting $a_0 = 1$, and then

$$\lambda_1 = \phi'(0) = a_1/b_2 \quad \text{and} \quad \lambda_2 = \phi'(\infty) = b_1,$$

so

$$\phi^f(z) = \frac{z^2 + b_2\lambda_1 z}{\lambda_2 z + b_2} \quad \text{with } b_2 \neq 0.$$

Then $b_2^{-1}\phi^f(b_2z)$ has the desired form. Further,

$$\mathrm{Resultant}(z^2+\lambda_1z, \lambda_2z+1) = \det\begin{vmatrix}1 & \lambda_1 & 0 & 0\\ 0 & 1 & \lambda_1 & 0\\ 0 & \lambda_2 & 1 & 0\\ 0 & 0 & \lambda_2 & 1\end{vmatrix} = 1 - \lambda_1\lambda_2,$$

which completes the proof of (i).

(ii) The proof is similar. We move α_1 to ∞, so

$$\phi(z) = \frac{a_0 z^2 + a_1 z + a_2}{b_1 z + b_2} \quad \text{with } a_0 \neq 0 \text{ and } \lambda_1 = \lambda_\phi(\infty) = \frac{b_1}{a_0}.$$

Then

$$\lambda_1\lambda_2 = 1 \implies \lambda_1 = \lambda_2 = 1 \implies b_1 = a_0.$$

Now the conjugations

$$z \to z - b_2 \quad \text{and} \quad z \to \sqrt{a_2 z}$$

yield

$$\phi(z) = z + a_1 + \frac{1}{z}.$$

We leave it as an exercise to compute a_1 in terms of λ_3. □

Resuming the proof of Theorem 2.45, we first verify that $\boldsymbol{s}$ is bijective on points. To check injectivity, suppose that $\phi_1, \phi_2 \in \mathrm{Hom}_2$ satisfy

$$\boldsymbol{s}(\phi_1) = \boldsymbol{s}(\phi_2).$$

Then $\sigma_3 = \sigma_1 - 2$ shows that

$$\sigma_i(\phi_1) = \sigma_i(\phi_2) \quad \text{for } i = 1, 2, 3.$$

The multipliers of a map ϕ are the roots of $\sum \sigma_i(\phi)T^{d-i}$, so the multiplier spectra of ϕ_1 and ϕ_2 are the same,

$$\Lambda_1(\phi_1) = \Lambda_1(\phi_2).$$

If $\lambda_1\lambda_2 \neq 1$, then Lemma 2.46(i) says that ϕ_1 and ϕ_2 are both $\mathrm{SL}_2(\mathbb{C})$-conjugate to $(z^2 + \lambda_1 z)/(\lambda_2 z + 1)$, so $\langle\phi_1\rangle = \langle\phi_2\rangle$ in M_2. The same is true if $\lambda_1\lambda_3 \neq 1$ or $\lambda_2\lambda_3 \neq 1$. Using the identities (2.9), (2.10), and (2.11), we are reduced to the case $\lambda_1 = \lambda_2 = \lambda_3 = 1$. But then Lemma 2.46(ii) says that ϕ_1 and ϕ_2 are both conjugate to $z + 1/z$.

Next we check surjectivity. Let $(s_1, s_2) \in \mathbb{A}^2(\mathbb{C})$, set $s_3 = s_1 - 2$, and define $\lambda_1, \lambda_2, \lambda_3$ by

$$T^3 + s_1 T^2 + s_2 T + s_3 = (T + \lambda_1)(T + \lambda_2)(T + \lambda_3).$$

Since

$$s_3 = s_1 - 2 \implies \lambda_1 + \lambda_2 + \lambda_3 = s_1 = s_3 + 2 = \lambda_1\lambda_2\lambda_3 + 2,$$

we obtain the usual formal identities (2.9), (2.10), and (2.11) for $\lambda_1, \lambda_2, \lambda_3$.

Suppose first that some $\lambda_i \neq 1$. Relabeling, we may assume that $\lambda_1 \neq 1$. Then

$$\begin{aligned} \lambda_1 \neq 1 &\implies \lambda_1\lambda_2 \neq 1 \\ &\implies \phi(z) = \frac{z^2 + \lambda_1 z}{\lambda_2 z + 1} \in \mathrm{Hom}_2, \end{aligned}$$

since

$$\mathrm{Resultant}\left(\frac{z^2 + \lambda_1 z}{\lambda_2 z + 1}\right) = 1 - \lambda_1\lambda_2 \neq 0.$$

The fixed points of ϕ are

$$\alpha_1 = 0, \qquad \alpha_2 = \infty, \qquad \alpha_3 = \frac{1 - \lambda_1}{1 - \lambda_2},$$

and one can check (using the identities) that

$$\lambda_\phi(\alpha_1) = \lambda_1, \quad \lambda_\phi(\alpha_2) = \lambda_2, \quad \lambda_\phi(\alpha_3) = \lambda_3.$$

Hence $\boldsymbol{s}(\phi) = (s_1, s_2)$, which proves that (s_1, s_2) is in the image of $\boldsymbol{s}$.

Finally, suppose that $\lambda_1 = \lambda_2 = \lambda_3 = 1$. Then we observe that

$$\lambda_1 = \lambda_2 = \lambda_3 = 1 \implies s_1 = s_2 = 3,$$

and

$$\phi(z) = z + z^{-1} \text{ satisfies } \Lambda_1(\phi) = \{1, 1, 1\},$$

so $\boldsymbol{s}(\phi) = (s_1, s_2)$.

We have shown that

$$\boldsymbol{s} = (\sigma_1, \sigma_2) \colon \mathsf{M}_2(\mathbb{C}) \longrightarrow \mathbb{A}^2(\mathbb{C})$$

is a bijective morphism, but that's not quite enough to guarantee that it is an isomorphism. For example, the standard map from $\mathbb{A}^1$ to a cuspidal cubic,

$$\mathbb{A}^1 \longrightarrow \{(x, y) \in \mathbb{A}^2 : y^2 = x^3\}, \quad t \longmapsto (t^2, t^3),$$

is a bijective morphism that is not an isomorphism. We sketch the remaining steps needed to prove that the map $\mathsf{M}_2 \to \mathbb{A}^2$ is an isomorphism.

Step 1. Show that $\boldsymbol{s}$ is proper. This follows from the valuative criterion, since every fiber $\boldsymbol{s}^{-1}(t)$ is complete, since it consists of a single point.

Step 2. Show that $\boldsymbol{s}$ is a finite morphism. We know from Step 1 that $\boldsymbol{s}$ is proper, and we are given that it is quasi-finite (has finite fibers). Then it is a general result that a proper quasi-finite map is finite.

Step 3. Use another general result which says that a finite map that is bijective on points and whose image is *nonsingular* is an isomorphism. □

Our general GIT construction (Theorem 2.24) says that M_d is contained in two larger spaces,

$$\mathsf{M}_d \subset (\mathsf{M}_d)^{\mathrm{s}} \subset (\mathsf{M}_d)^{\mathrm{ss}}.$$

For $d = 2$, the stable and semistable quotients coincide, so we denote them by $\overline{\mathsf{M}}_2$. Theorem 2.45 tells us that $\mathbb{A}^2 \cong \mathsf{M}_2$, which gives an inclusion

$$\mathbb{A}^2 \cong \mathsf{M}_2 \subset \overline{\mathsf{M}}_2.$$

The space $\overline{\mathsf{M}}_2$ is complete, and the points of

$$\overline{\mathsf{M}}_2 \smallsetminus \mathsf{M}_2$$

correspond to degenerate maps of degree 2.

Theorem 2.47 (Milnor [78] over $\mathbb{C}$, Silverman [109] over $\mathbb{Z}$). *The isomorphism*

$$\boldsymbol{s} \colon \mathsf{M}_2 \longrightarrow \mathbb{A}^2$$

described in Theorem 2.45 *extends to an isomorphism*

$$\bar{\boldsymbol{s}} \colon \overline{\mathsf{M}}_2 \longrightarrow \mathbb{P}^2$$

such that the following diagram commutes:

$$\begin{array}{ccc} \mathsf{M}_2 & \xrightarrow[\boldsymbol{s}]{\sim} & \mathbb{A}^2 \\ \big\downarrow & & \big\downarrow {\scriptstyle (x,y) \mapsto [x,y,1]} \\ \overline{\mathsf{M}}_2 & \xrightarrow[\bar{\boldsymbol{s}}]{\sim} & \mathbb{P}^2. \end{array}$$

The boundary $\overline{\mathsf{M}}_2 \smallsetminus \mathsf{M}_2$ *may be described as follows. For* $[a,b] \in \mathbb{P}^1$, *let*

$$\phi_{a,b}(z) = \frac{az}{bz+b} = [0,a,0,0,b,b] \in (\mathrm{Hom}_2)^{\mathrm{s}} \subset \mathbb{P}^5.$$

Then the map

$$\mathbb{P}^1 \longrightarrow \overline{\mathsf{M}}_2 \smallsetminus \mathsf{M}_2 \cong \mathbb{P}^1, \qquad [a,b] \longmapsto \langle \phi_{a,b} \rangle,$$

is a double cover determined by the equivalence

$$\langle \phi_{a,b} \rangle = \langle \phi_{b,a} \rangle.$$

We can interpret Theorem 2.47 as follows. Suppose that $(\phi_t)_{t\neq 0}$ is a family of maps of degree 2 that cannot be filled in at $t=0$, i.e., we have a morphism

$$\mathbb{A}^1 \smallsetminus \{0\} \longrightarrow \mathrm{Hom}_2, \quad t \longmapsto \phi_t,$$

that does not extend to a morphism on $\mathbb{A}^1$. Then there is a change of variables $f \in \mathrm{PGL}_2$ and pair $[a,b] \in \mathbb{P}^1$ such that

$$\lim_{t\to 0} \phi_t^f = \phi_{a,b}.$$

Further, $[a,b]$ is uniquely determined up to reversing a and b.

Remark 2.48. We have used the maps $\phi_{a,b}$ because they fit naturally into the normal forms lemma as the multipliers go to infinity. As an alternative, we can conjugate by z^{-1} to cover the boundary of M_2 with polynomial maps

$$\frac{1}{\phi_{a,b}(z^{-1})} = \frac{bz+b}{a} = [0,b,b,0,0,a].$$

In any case, the boundary $\overline{\mathsf{M}}_2 \smallsetminus \mathsf{M}_2$ consists of an $\mathbb{A}^1$-family of linear maps corresponding to $[a,b] \neq [1,0]$, together with a single constant map corresponding to $[a,b]=[1,0]$.

2.6. Uniform bounds for $\mathrm{Aut}(\phi)$

In this section we consider the automorphism group of a rational map ϕ, which is the group of linear maps f satisfying $\phi^f = \phi$. This group is analogous to the group of automorphisms of an abelian variety, with the difference that the automorphism group of an abelian variety is always nontrivial, since it contains the map $P \to -P$. By way of contrast, most (but not all) rational maps ϕ have a trivial automorphism group.

Definition 2.49. Let $\phi \in \mathrm{Hom}_d^n$. The *automorphism group* of ϕ is the group

$$\mathrm{Aut}(\phi) = \{f \in \mathrm{PGL}_{n+1} : \phi^f = \phi\}.$$

Alternative names for the automorphism group that appear in the literature include *stabilizer group* and the *group of self-similarities*.

Remark 2.50. The existence of maps with $\mathrm{Aut}(\phi) \neq 1$ prevents the existence of a universal family of maps over M_d^n, i.e., we cannot make M_d^n into a fine moduli space.

Proposition 2.51 (Petsche, Szpiro, Tepper [95]). $\mathrm{Aut}(\phi)$ *is a finite group.*

PROOF SKETCH. Let $\mathrm{PrePer}^{(m)}(\phi)$ denote the set of preperiodic points for ϕ whose orbit consists of at most m points. We observe that if $f \in \mathrm{Aut}(\phi)$ and $P \in \mathrm{PrePer}^{(m)}(\phi)$, then

$$\mathcal{O}_\phi(P) \xrightarrow[\sim]{f^{-1}} \mathcal{O}_{\phi^f}\big(f^{-1}(P)\big) = \mathcal{O}_\phi\big(f^{-1}(P)\big),$$

so $f^{-1}(P) \in \mathrm{PrePer}^{(m)}(\phi)$. Hence an automorphism $f \in \mathrm{Aut}(\phi)$ induces a permutation

$$f\colon \mathrm{PrePer}^{(m)}(\phi) \to \mathrm{PrePer}^{(m)}(\phi).$$

This gives a homomorphism

$$\mathrm{Aut}(\phi) \longrightarrow \mathrm{Sym}\big(\mathrm{PrePer}^{(m)}(\phi)\big). \tag{2.12}$$

Suppose that we could show that $\mathrm{PrePer}(\phi)$ is Zariski dense in $\mathbb{P}^n$. Then we can find an m such that $\mathrm{PrePer}^{(m)}(\phi)$ contains at least $n+2$ points in general position in $\mathbb{P}^n$. Any $f \in \mathrm{PGL}_{n+1}$ fixing $n+2$ points in general position is the identity map, so (2.12) is an injective homomorphism of $\mathrm{Aut}(\phi)$ into a finite group.

To see that $\mathrm{PrePer}(\phi)$ is Zariski dense, let Z be its Zariski closure, and write $Z = Z_1 \cup \cdots \cup Z_r$ as a union of irreducible components. Then ϕ permutes the Z_i, so replacing ϕ by some iterate, we may assume that $\phi(Z_1) = Z_1 = \phi^{-1}(Z_1)$. Then comparing the degrees of

$$\phi\colon \mathbb{P}^n \longrightarrow \mathbb{P}^n \quad \text{and} \quad \phi|_{Z_1}\colon Z_1 \longrightarrow Z_1$$

gives

$$d^{\dim(Z_1)} = \deg\big(\phi|_{Z_1}\big) = \deg(\phi) = d^n.$$

Hence $\dim(Z_1) = \dim(\mathbb{P}^n)$, and since Z_1 is a closed subset of $\mathbb{P}^n$, we must have $Z_1 = \mathbb{P}^n$. (For further details, see [95, Proposition 2].) □

Remark 2.52. While proving Proposition 2.51, we sketched the proof that $\mathrm{PrePer}(\phi)$ is Zariski dense in $\mathbb{P}^n$. It is in fact true, but more difficult to prove, that $\mathrm{Per}(\phi)$ is Zariski dense in $\mathbb{P}^n$. See [22] for the result over $\mathbb{C}$ and [40] for a general proof using model theory.

Proposition 2.51 says that $\mathrm{Aut}(\phi)$ is finite, but for some applications it is important to have a uniform bound for the size of $\mathrm{Aut}(\phi)$, as in the next result.

Theorem 2.53 (Levy [66]). *Let $n \ge 1$ and $d \ge 2$. Then there is a constant $C(d,n)$ such that*

$$\#\mathrm{Aut}(\phi) \le C(d,n) \quad \text{for all } \phi \in \mathrm{Hom}_d^n.$$

PROOF SKETCH (char 0). Step 1. Show that every $A \in \mathrm{Aut}(\phi)$ is diagonalizable, i.e., there exists a $B \in \mathrm{PGL}_{n+1}$ such that $B^{-1}AB$ is diagonal. The proof is a detailed analysis of how a nontrivial Jordan block would have to act on Hom_d^n.

Step 2. From Step 1, every *abelian* subgroup of $\mathrm{Aut}(\phi)$ is simultaneously diagonalizable. Next prove that every subgroup of $\mathrm{Aut}(\phi)$ consisting entirely of diagonal matrices has order at most d^{n+1}. This is done by examining the action of the matrices on the x_i^d terms appearing in the coordinate functions of $\phi = [\phi_0, \dots, \phi_n]$.

Step 3. Now complete the proof by using a result of G. A. Miller (1906): The size of a finite group is bounded in terms of the size of its largest abelian subgroup. □

Example 2.54. It is a classical result that in characteristic 0, the only finite subgroups of PGL_2 are $\mathfrak{C}_k$, $\mathfrak{D}_{2k}$, $\mathfrak{A}_4$, $\mathfrak{A}_5$, and $\mathfrak{S}_4$. (These are, respectively, a cyclic group of order k, a dihedral group of order $2k$, alternating groups on 4 and 5 letters, and the symmetric group on 4 letters.)It is also not hard to show that

$$\mathfrak{C}_k \subset \mathrm{Aut}(\phi) \iff \phi^f \sim z\psi(z^k) \text{ for some } \psi \text{ and } f \in \mathrm{PGL}_2.$$

(See, e.g., [107, Proposition 7.2].) In particular, since $\mathfrak{D}_{2k}$ contains a copy of $\mathfrak{C}_k$, we have

$$\begin{aligned}\#\,\mathrm{Aut}(\phi) > 60 = \#\,\mathfrak{A}_5 &\implies \mathrm{Aut}(\phi) = \mathfrak{C}_k \text{ or } \mathfrak{D}_{2k},\\ &\implies \deg\phi = \deg z\psi(z^k) \ge k-1 \ge \frac{1}{2}\,\#\,\mathrm{Aut}(\phi) - 1.\end{aligned}$$

This gives an explicit value for the constant in Theorem 2.53 when $n = 1$,

$$\#\,\mathrm{Aut}(\phi) \le \max\{60, 2\deg\phi + 2\}.$$

Corollary 2.55 (Levy [66]). *Most maps $\phi \in \mathsf{M}_d^n$ have trivial automorphism group. In other words,*

$$\{\phi \in \mathsf{M}_d^n : \mathrm{Aut}(\phi) \ne 1\}$$

is a finite union of proper subvarieties of M_d^n.

PROOF SKETCH. From Theorem 2.53, there are only finitely many possibilities for $\mathrm{Aut}(\phi)$ as an abstract group, so it suffices to look at

$$\{\phi \in \mathsf{M}_d^n : \mathrm{Aut}(\phi) \cong G\}$$

for a fixed finite group G. Then one shows that such ϕ fall into finitely many algebraic families. For details, see [66, Corollary 3.6]. □

2.7. Rationality of M_d^1

We have seen that $\mathsf{M}_2^1 = \mathbb{A}^2$. Further, it is clear that all of the moduli spaces M_d^n are unirational, i.e., they admit a dominant rational map $\mathbb{P}^L \dashrightarrow \mathsf{M}_d^n$. This is obvious, because $\mathrm{Hom}_d^n \subset \mathbb{P}^N$ is a rational variety. It is less clear whether M_d^n is itself rational, but at least for maps of $\mathbb{P}^1$ this is true.

Theorem 2.56 (Levy [66]). *M_d^1 is a rational variety.*

PROOF. More precisely, let

$$\mathcal{P}_{d+1} = (\text{space of unmarked sets of } d+1 \text{ points on } \mathbb{P}^1).$$

It is an open subset of

$$\mathcal{P}_{d+1} \subset (\mathbb{P}^1)^{d+1}/\mathcal{S}_{d+1}\,\mathrm{PGL}_2,$$

where $\mathcal{S}_{d+1}$ permutes the coordinates and PGL_2 acts diagonally. (Of course, we really only take the quotient of the stable locus.) It is known that

$$\mathcal{P}_{d+1} \text{ is a rational variety.}$$

To show that M_d is a rational variety, we show that it is birational to a vector bundle over $\mathcal{P}_{d+1}$.

Consider the map

$$\mathrm{Fix}\colon \mathrm{Hom}_d \longrightarrow (\mathbb{P}^1)^{d+1}/\mathcal{S}_{d+1}, \qquad \phi \longmapsto \mathrm{Fix}(\phi).$$

This is certainly a rational map (actually it's a morphism).

Claim 1. Fix *is surjective.*

Claim 2. Fix *has rational fibers.*

We begin with Claim 2 and consider a fiber of the map Fix. Let

$$\boldsymbol{P} = (P_1, \dots, P_{d+1}) \in (\mathbb{P}^1)^{d+1}$$

be a generic point of $(\mathbb{P}^1)^{d+1}$. Writing

$$P_i = [x_i, y_i] \quad \text{and} \phi_{a,b} = F_a(x,y)/F_b(x,y),$$

we have

$$\operatorname{Fix}^{-1}(\boldsymbol{P}) = \bigl\{\phi_{a,b} : y_i F_a(x_i, y_i) = x_i F_b(x_i, y_i) \text{ for all } 1 \le i \le d+1\bigr\}.$$

Each equation

$$y_i F_a(x_i, y_i) = x_i F_b(x_i, y_i)$$

imposes one homogeneous linear condition on the $2d+2$ homogeneous coordinates of $[a, b]$, and since the points $P_1, \dots, P_{d+1}$ are generic, these linear conditions are independent. Hence $\operatorname{Fix}^{-1}(\boldsymbol{P})$ is a linear subspace of $\mathbb{P}^{2d+1}$ of codimension $d+1$. In particular, $\operatorname{Fix}^{-1}(\boldsymbol{P})$ is rational. This proves Claim 2.

Next we need to show that there is at least one point $[a, b] \in \operatorname{Fix}^{-1}(\boldsymbol{P})$ such that $[F_a, F_b]$ is a morphism of degree d, i.e., we need to find a point $\phi \in \operatorname{Hom}_d$ that maps to $\boldsymbol{P}$. Let

$$R(X,Y) = \begin{pmatrix} \text{homogeneous polynomial of degree } d+1 \\ \text{with roots } P_1, \dots, P_{d+1}. \end{pmatrix}.$$

We need to find homogeneous polynomials $F_a(X,Y)$ and $F_b(X,Y)$ of degree d satisfying

$$R = YF_a - XF_b \quad \text{and} \quad \gcd(F_a, F_b) = 1.$$

(The first condition forces ϕ to have the correct fixed points, and the second condition forces ϕ to have the correct degree.) Conjugating, we may assume without loss of generality that no P_i is $[0,1]$ or $[1,0]$, so

$$R(X,Y) = \alpha X^{d+1} + \cdots + \beta Y^{d+1} \quad \text{with } \alpha\beta \ne 0.$$

Take

$$F_b = -\alpha X^d \qquad \text{and} \qquad F_a = \frac{R + XF_b}{Y} = \frac{R - \alpha X^{d+1}}{Y} = \cdots + \beta Y^d.$$

Then

$$F_b(X,Y) = 0 \iff X = 0 \quad \text{and} \quad F_a(0,1) = \beta \ne 0.$$

Hence $\phi = F_a/F_b \in \operatorname{Hom}_d$, which proves that

$$\operatorname{Fix}\colon \operatorname{Hom}_d \longrightarrow (\mathbb{P}^1)^{d+1}/\mathcal{S}_{d+1}$$

is surjective.

The map Fix descends to define a dominant rational map $\mathcal{F}$ on M_d that fits into the commutative diagram

$$\begin{array}{ccc} \operatorname{Hom}_d & \xrightarrow{\ \operatorname{Fix}\ } & (\mathbb{P}^1)^d/\mathcal{S}_d \\ \downarrow & & \downarrow \\ \mathsf{M}_d & \xrightarrow{\ \mathcal{F}\ } & \mathcal{P}_{d+1} \subset (\mathbb{P}^1)^{d+1}/\mathcal{S}_{d+1}\operatorname{PGL}_2. \end{array}$$

We claim that the generic fiber of $\mathcal{F}$ is rational. Let

$$\operatorname{Aut}(\boldsymbol{P}) = \bigl\{\gamma \in \operatorname{PGL}_2 : \gamma(\boldsymbol{P}) = \boldsymbol{P}\bigr\},$$

i.e., $\operatorname{Aut}(\boldsymbol{P})$ is the set of γ that permute the $d+1$ points in the set $\boldsymbol{P}$. Then

$$\mathcal{F}^{-1}(\boldsymbol{P}) \cong \operatorname{Fix}^{-1}(\boldsymbol{P}) / \operatorname{Aut}(\boldsymbol{P}).$$

We know from above that $\operatorname{Fix}^{-1}(\boldsymbol{P})$ is rational, so if $\operatorname{Aut}(\boldsymbol{P}) = 1$, then we're done. The next lemma describes the possible values of $\operatorname{Aut}(\boldsymbol{P})$.

Lemma 2.57. *Let $\boldsymbol{P}$ be a generic point of $(\mathbb{P}^1)^{d+1}$. Then*

$$\operatorname{Aut}(\boldsymbol{P}) = \begin{cases} 1 & \text{if } d \geq 4, \\ \mathbb{Z}/2\mathbb{Z} \times \mathbb{Z}/2\mathbb{Z} & \text{if } d = 3, \\ \mathcal{S}_3 & \text{if } d = 2. \end{cases}$$

PROOF. Let

$$\sigma \in \operatorname{Aut}(\boldsymbol{P}),$$

so $\sigma \in \mathrm{PGL}_2$ permutes $P_1, \ldots, P_{d+1}$. Suppose that there is a k-cycle,

$$P_{i_1} \to P_{i_2} \to P_{i_3} \to \cdots \to P_{i_k} \to P_{i_1},$$

with $k \geq 4$. Changing coordinates, we may assume the k-cycle has the form

$$\mu \xrightarrow{\sigma} \infty \xrightarrow{\sigma} 0 \xrightarrow{\sigma} 1 \xrightarrow{\sigma} \lambda,$$

where $\lambda, \mu \notin \{0, 1, \infty\}$. Writing

$$\sigma = \frac{az+b}{cz+d},$$

the condition $\sigma(\infty) = 0$ forces $a = 0$. This means we can take $b = 1$, and then

$$\sigma(0) = 1 \text{ forces } d = b = 1, \quad \text{so } \sigma = \frac{1}{cz+1}.$$

From this we see that

$$\lambda = \sigma(1) = \frac{1}{c+1} \quad \text{and} \quad \mu = \sigma^{-1}(\infty) = -\frac{1}{c},$$

and hence

$$\lambda = \frac{\mu}{1-\mu}.$$

This contradicts the assumption that $\boldsymbol{P}$ is generic, which proves:

All cycles of σ acting on $\boldsymbol{P}$ have length 1, 2, or 3.

Suppose that σ contains a 3-cycle. Changing coordinates, the 3-cycle has the form

$$0 \longrightarrow 1 \longrightarrow \infty \longrightarrow 0,$$

which forces

$$\sigma = \frac{1}{1-z}.$$

If $d \geq 3$, then $\boldsymbol{P}$ contains at least four points, say $\boldsymbol{P} = \{0, 1, \infty, \lambda, \ldots\}$. But then it is forced to contain $\sigma(\lambda) = 1/(1-\lambda)$, which it won't since $\boldsymbol{P}$ is generic. Hence for $d \geq 3$, the map σ contains no 3-cycles.

Suppose that σ contains two 2-cycles. Conjugating, we may suppose that the 2-cycles look like

$$0 \longrightarrow \infty \longrightarrow 0 \quad \text{and} \quad 1 \longrightarrow \lambda \longrightarrow 1.$$

This forces

$$\sigma = \frac{\lambda}{z}.$$

If $d = 3$, this map σ stabilizes $\boldsymbol{P} = \{0, 1, \infty, \lambda\}$, while $d \geq 4$, then σ does not stabilize

$$\boldsymbol{P} = \{0, 1, \infty, \lambda, \mu, \ldots\} \qquad \text{for generic } \lambda, \mu.$$

Hence elements of $\operatorname{Aut}(\boldsymbol{P})$ with two 2-cycles occur only for $d = 3$.

Finally, suppose σ has one 2-cycle and that all of the other P_i are fixed by σ. If $d \geq 4$, then σ fixes 3 points, so $\sigma(z) = z$. For $d = 3$, after conjugation we may assume that $\boldsymbol{P} = \{0, 1, \infty, \lambda\}$ and

$$0 \longrightarrow 1 \longrightarrow 0, \quad \infty \longrightarrow \infty, \quad \text{and} \quad \lambda \longrightarrow \lambda.$$

This forces

$$\sigma(z) = 1 - z, \quad \text{which does not fix a generic } \lambda.$$

This completes the proof for $d \geq 3$. For $d = 2$, after a change of coordinates we have $\boldsymbol{P} = \{0, 1, \infty\}$, so $\operatorname{Aut}(\boldsymbol{P}) = \mathcal{S}_3$. □

We resume the proof of Theorem 2.56. Using Lemma 2.57, the proof of the Theorem 2.56 is complete for $d \geq 4$, since

$$\begin{aligned} d \geq 4 &\implies \operatorname{Aut}(\boldsymbol{P}) = 1 \\ &\implies \mathcal{F}^{-1}(\boldsymbol{P}) \cong \operatorname{Fix}^{-1}(\boldsymbol{P}) / \operatorname{Aut}(\boldsymbol{P}) = \operatorname{Fix}^{-1}(\boldsymbol{P}) \cong \mathbb{P}^d. \end{aligned}$$

(Here $\cong$ denotes that two varieties are birational to one another.) This shows that for $d \geq 4$, the map

$$\mathcal{F} \colon \mathsf{M}_d \longrightarrow \mathcal{P}_{d+1}$$

is a surjective rational map whose fibers are rational varieties. Since it is known that $\mathcal{P}_{d+1}$ is rational, this proves the M_d is rational.

For $d = 2$ and $d = 3$, the group $\operatorname{Aut}(\boldsymbol{P})$ is nontrivial, but quite small. In these cases, one can use a result of Noether and some calculations to prove that the quotient

$$\operatorname{Fix}^{-1}(\boldsymbol{P}) / \operatorname{Aut}(\boldsymbol{P})$$

is still rational. For details, see [66]. The proof depends heavily on the fact that $\operatorname{Aut}(\boldsymbol{P})$ is very small. (Of course, for $d = 2$ we already know from Theorem 2.45 that $\mathsf{M}_2 \cong \mathbb{A}^2$, which is stronger.) □

2.8. Special loci in M_d^n

There are many specific types of rational maps that play prominent roles in dynamics. It is an interesting problem to understand the loci of these maps in the appropriate moduli spaces. We give two examples of such loci in M_2, and then we pose a number of questions.

Proposition 2.58. *Under the identification $\mathsf{M}_2 \cong \mathbb{A}^2$, we have:*

(a) *The set of polynomial maps*

$$\bigl\{ \langle \phi \rangle \in \mathsf{M}_2 : \phi(z) = az^2 + bz + c \bigr\}$$

is the line $\sigma_1 = 2$. More precisely, every polynomial map of degree 2 is conjugate to a map of the form $z^2 + c$, and

$$\sigma_1(z^2 + c) = 2 \quad \textit{and} \quad \sigma_2(z^2 + c) = 4c.$$

(b) *The set of maps in* M_2 *with nontrivial automorphism group,*

$$\{\langle\phi\rangle \in \mathsf{M}_2 : \operatorname{Aut}(\phi) \neq 1\},$$

forms a cuspidal cubic curve in $\mathbb{A}^2 \cong \mathsf{M}_2$. *Every* ϕ *on this curve satisfies* $\operatorname{Aut}(\phi) = \mathbb{Z}/2\mathbb{Z}$ *except for the cusp, where* $\operatorname{Aut}(\phi) = \mathcal{S}_3$.

PROOF. The proof of (a) is easy. The proof of (b) is sketched later (Proposition 4.15), or see [109]. □

Definition 2.59. Let $\phi\colon \mathbb{P}^N \to \mathbb{P}^N$ be a rational map of degree $d \geq 2$.

(1) Let $\mathbb{A}^N = \mathbb{P}^N \smallsetminus \{x_0 = 0\}$. The map ϕ is an *affine morphism* if ϕ induces a morphism $\phi\colon \mathbb{A}^N \to \mathbb{A}^N$.

(2) The map ϕ is an *affine automorphism* if ϕ is an affine morphism that has an inverse that is also an affine morphism.

(3) The *locus of indeterminacy of* ϕ is the set

$$Z(\phi) = \{P \in \mathbb{P}^N : \phi \text{ is not defined at } P\}.$$

An affine automorphism ϕ is *regular* if

$$Z(\phi) \cap Z(\phi^{-1}) = \varnothing.$$

(4) The map ϕ is *algebraically stable* if

$$\deg(\phi^n) = (\deg\phi)^n \quad \text{for all } n \geq 1.$$

Here we use degree to mean the degree of the polynomials that define the map, i.e.,

$$\phi^*\mathcal{O}_{\mathbb{P}^N}(1) = \mathcal{O}_{\mathbb{P}^N}(\deg\phi).$$

So ϕ is algebraically stable if

$$(\phi^n)^*\mathcal{O}_{\mathbb{P}^N}(1) = \bigl(\phi^*\mathcal{O}_{\mathbb{P}^N}(1)\bigr)^{\otimes n} \quad \text{for all } n \geq 1.$$

Remark 2.60. It's not hard to prove that regular affine automorphisms are algebraically stable. When we discuss height functions in Section 5.7, we will see that regular affine automorphisms have many nice arithmetic properties.

Example 2.61. The Hénon map

$$\phi\colon \mathbb{A}^2 \longrightarrow \mathbb{A}^2, \quad \phi(x,y) = (y + a + bx^2, cx), \quad \text{with } bc \neq 0,$$

is a regular affine automorphism of degree 2. Hénon maps were originally studied over $\mathbb{R}^2$ because they are chaotic and may have strange attractors, depending on the values of a, b, c.

In terms of projective coordinates, the Hénon map is given by

$$\phi\colon \mathbb{P}^2 \longrightarrow \mathbb{P}^2, \quad \phi\bigl([x,y,z]\bigr) = [yz + az^2 + bx^2, cxz, z^2].$$

Let $i + j + k = 0$ and consider the effect of the diagonal 1-parameter subgroup

$$\ell_{ijk}(\alpha) = \begin{pmatrix} \alpha^i & 0 & 0 \\ 0 & \alpha^j & 0 \\ 0 & 0 & \alpha^k \end{pmatrix}, \quad \text{i.e., } x \mapsto \alpha^i x,\ y \mapsto \alpha^j y,\ z \mapsto \alpha^k z.$$

There are 6 monomials of degree 2 in the variables x, y, z, so Hom_2^2 sits inside $\mathbb{P}^{17}$. The Hénon map ϕ corresponds to the point in $\operatorname{Hom}_2^2 \subset \mathbb{P}^{17}$ with coordinates

$$\begin{array}{cccccccccc} & yz & z^2 & x^2 & xz & z^2 & & & & \\ & \downarrow & \downarrow & \downarrow & \downarrow & \downarrow & & & & \\ [& 1, & a, & b, & c, & 1, & 0, & 0, & \dots &]. \end{array}$$

The 1-parameter subgroup ℓ_{ijk} acts on ϕ by sending it to

$$\ell_{ijk}(\alpha)\cdot[1,a,b,c,1,0,0,\dots] = [\alpha^{j+k},\alpha^{2k}a,\alpha^{2i}b,\alpha^{i+k}c,\alpha^{2k},0,0,\dots].$$

Since $bc \ne 0$, all of the terms α^{j+k}, α^{2i}, α^{i+k}, and α^{2k} appear, so

$$\mu(\phi,\ell_{ijk}) = \max\{-j-k,-2i,-i-k,-2k\}.$$

We have complete freedom in choosing i,j,k subject to the condition $i+j+k=0$, so substituting $k=-i-j$, we find that

$$\mu(\phi,\ell_{ijk}) = \max\{i,-2i,j,2i+2j\}.$$

Since one of i and $-2i$ is nonnegative, we have $\mu(\phi,\ell_{ijk}) \ge 0$, while taking $i=0$ and $j<0$ gives $\mu(\phi,\ell_{0jk})=0$. This shows that $\langle\phi\rangle$ is not in the stable locus $(\mathsf{M}_2^2)^{\mathrm{s}}$. The results from the diagonal 1-parameter subgroups ℓ_{ijk} suggest that $\langle\phi\rangle$ is in the semistable locus $(\mathsf{M}_2^2)^{\mathrm{ss}}$, but a complete verification would require showing that $\mu(\phi,\ell)\ge 0$ for all 1-parameter subgroups, not just the diagonal subgroups.

Question 2.62. We pose some questions related to stability conditions and moduli spaces.

(1) Which affine morphisms are stable? semistable? Describe the locus of affine morphisms in M_d^N, $(\mathsf{M}_d^N)^{\mathrm{s}}$, and $(\mathsf{M}_d^N)^{\mathrm{ss}}$. For $N=1$ and $d=2$, affine morphisms are polynomial maps and we saw in Proposition 2.58(a) that they form a line in $\mathsf{M}_2^1 \cong \mathbb{A}^2$.

(2) Same questions for affine automorphisms. Note that in this case the question is vacuous unless $n \ge 2$, since we always assume that $d \ge 2$.

(3) Same questions for regular affine automorphisms. As a warm-up, what is the locus of the (semistable) Hénon maps in $(\mathsf{M}_2^2)^{\mathrm{ss}}$?

(4) What is the relationship (if any) between algebraic stability and GIT-stability or semistability? In particular, what is the locus of algebraically stable maps in $(\mathsf{M}_d^N)^{\mathrm{ss}}$? (Note that this locus includes all of M_d^N, since morphisms are algebraically stable.)

Remark 2.63. The relation of "algebraic stability" to "GIT moduli-theoretic stability" is not clear. We remark that there are a many other kinds of geometrically defined stability that are used in complex dynamics, including for example *topological stability*, *qc-stability*, *postcritical stability*, and *J-stability*. Thus ϕ is topologically stable if there is a neighborhood U of ϕ in Hom_d^n such that for all $\psi \in U$ there is a homeomorphism f of $\mathbb{P}^n(\mathbb{C})$ such that $\phi = \psi^f$. Similarly, the map ϕ is quasi-conformally (qc)-stable if it is topologically stable and the conjugating map f can be taken to be quasi-conformal.[2] The map ϕ is postcritically stable if all critical point relations $\phi^i(c) = \phi^j(c')$ persist in a neighborhood of ϕ. And ϕ is J-stable if the map

$$\phi \longmapsto \mathrm{Julia}(\phi)$$

is continuous in a neighborhood of ϕ, where we use the Hausdorff topology[3] on the set of compact subsets of $\mathbb{P}^n(\mathbb{C})$.

For maps of $\mathbb{P}^1$, a fair amount is known about these various types of stability, including the following.

[2]A map on $\mathbb{C}$ is quasi-conformal if it sends infinitesimal circles to infinitesimal ellipses whose eccentricity is uniformly bounded above and below.

[3]In the Hausdorff topology, two compact sets are close if they are each contained in an ϵ-neighborhood of the other.

Theorem 2.64 (McMullen–Sullivan [76]; see also Mañé–Sad–Sullivan [67] and McMullen [74]). *For maps $\phi \in \mathrm{Hom}_d(\mathbb{C})$,*

$$\begin{aligned} \phi \text{ is topologically stable} &\iff \phi \text{ is qc-stable} \\ &\iff \phi \text{ is postcritically stable} \\ &\iff \phi \text{ is } J\text{-stable.} \end{aligned}$$

Remark 2.65. Polynomial maps of one variable have been much studied. The locus of such maps in M_d has a simple structure, which we now describe. In general, a rational map $\phi\colon \mathbb{P}^1 \to \mathbb{P}^1$ is a *polynomial map* if it has a totally ramified fixed point. Moving the totally ramified fixed point to ∞, we can write ϕ as a polynomial in the variable z. The only change of variables $f \in \mathrm{PGL}_2$ that preserve the point at ∞ are affine linear maps $f(z) = az + b$. Let

$$\phi(z) = c_0 z^d + c_1 z^{d-1} + c_2 z^{d-2} + \cdots + c_d \quad \text{with } c_0 \neq 0.$$

Assuming that the characteristic does not divide d, we can make the change of variables

$$f(z) = \frac{z}{c_0^{1/(d-1)}} - \frac{c_1}{dc_0}$$

to put $\phi(z)$ in the form

$$\phi(z) = z^d + c_2 z^{d-2} + \cdots + c_d \in K[z].$$

The only elements of PGL_2 that preserve this form are are the maps $f(z) = \zeta z$ with $\zeta \in \boldsymbol{\mu}_{d-1}$, and the action of f on ϕ is

$$\phi^f(z) = z^d + \zeta^{d-3} c_2 z^{d-2} + \zeta^{d-4} c_3 z^{d-3} + \cdots + \zeta c_{d-2} z^2 + c_{d-1} z + \zeta^{-1} c_d.$$

So if we define an action of $\boldsymbol{\mu}_{d-1}$ on $\mathbb{A}^{d-1}$ by

$$\zeta * (x_2, x_3, \ldots, x_{d-2}, x_{d-1}, x_d) = (\zeta^{d-3} x_2, \zeta^{d-4} x_3, \ldots, \zeta x_{d-2}, x_{d-1}, \zeta^{-1} x_d),$$

then there is a natural isomorphism

$$\begin{aligned} \mathbb{A}^{d-1}/\boldsymbol{\mu}_{d-1} &\xrightarrow{\ \sim\ } \{\text{polynomial maps in } \mathsf{M}_d\}, \\ (c_2, \ldots, c_d) &\longmapsto \langle z^d + c_d z^{d-2} + \cdots + c_2 \rangle, \end{aligned}$$

valid in characteristics not dividing d. The quotient is a (singular) variety, since there are points in $\mathbb{A}^{d-1}$ that have nontrivial stabilizer. For example, the polynomial $z^d + cz$ is fixed by all of $\boldsymbol{\mu}_{d-1}$. There may also be higher dimensional subvarieties with nontrivial stabilizer. For example, if d is odd, then the automorphism group of every polynomial of the form

$$z^d + c_2 z^{d-2} + c_4 z^{d-4} + \cdots + c_{d-3} z^3 + c_{d-1} z$$

contains the map $f(z) = -z$.

Remark 2.66. The moduli space M_2 has a very simple structure, it is isomorphic to $\mathbb{A}^2$. For larger values of d, the space M_d is rational (Theorem 2.56), but it has higher dimension and its finer structure appears to be quite complicated. The orbits of the critical points are an important determinant of the complexity of the dynamics of a rational map $\phi(z)$, so maps with fewer critical points are somehow less complicated. The extreme case is the set of polynomial maps of degree two.

These are maps with only two critical points, one of which is fixed and totally ramified. More generally, M_2 is a simpler space than M_d for $d \ge 3$ because rational maps of degree two have only two critical points. This suggests looking at the space[4]

$$\mathrm{BiCrit}_d = \{\phi \in \mathrm{Hom}_d : \phi \text{ has exactly two critical points}\}.$$

Thus $\mathrm{BiCrit}_2 = \mathrm{Hom}_2$, but in general the set BiCrit_d is a proper subset of Hom_d. We note that the Hurwitz genus formula

$$2d - 2 = \sum_{P \in \mathbb{P}^1} e_P(\phi) - 1$$

implies that every ϕ has at least two critical points, since the ramification indices satisfy $1 \le e_P(\phi) \le d$.

It is not hard to show that BiCrit_d is an algebraic subset of Hom_d of dimension 5 and that PGL_2 acts on BiCrit_d via conjugation in the usual way. Hence the geometric quotient

$$\mathsf{M}_d^{\mathrm{BiCrit}} = \mathrm{BiCrit}_d \,/\, \mathrm{PGL}_2$$

exists. Generalizing Theorem 2.45, one finds that

$$\mathsf{M}_d^{\mathrm{BiCrit}} \cong \mathbb{A}^2.$$

More precisely, let $\phi \in \mathrm{BiCrit}_d$ and use a linear fractional transformation to move the two critical points to 0 and ∞. Then ϕ necessarily has the form

$$\phi(z) = \frac{\alpha z^d + \beta}{\gamma z^d + \delta} \quad \text{with } \alpha\delta - \beta\gamma \ne 0,$$

and one can show that the quantities

$$A(\phi) = \frac{\beta\gamma}{\alpha\delta - \beta\gamma}, \quad B(\phi) = \frac{\alpha^{d+1}\beta^{d-1}}{(\alpha\delta - \beta\gamma)^d}, \quad C(\phi) = \frac{\gamma^{d-1}\delta^{d+1}}{(\alpha\delta - \beta\gamma)^d},$$

depend only on the conjugacy class of ϕ and satisfy the single relation

$$BC = A^{d-1}(A+1)^{d+1}.$$

Further, working over an algebraically closed field, the values of $A(\phi)$, $B(\phi)$, and $C(\phi)$ determine the PGL_2-conjugacy class of ϕ. Hence

$$(A, B, C)\colon \mathsf{M}_d^{\mathrm{BiCrit}} \longrightarrow \mathbb{A}^3$$

maps $\mathsf{M}_d^{\mathrm{BiCrit}}$ onto the variety

$$\{(x, y, z) \in \mathbb{A}^3 : yz = x^{d-1}(x+1)^{d+1}\}, \tag{2.13}$$

and this map is a bijection on points. Notice that the variety (2.13) is singular at $(0,0,0)$ and $(-1,0,0)$, which are the points corresponding respectively to the bicritical maps $\phi(z) = z^d$ and $\phi(z) = z^{-d}$ having nontrivial automorphism groups.For proofs and further information about the geometry and topology of BiCrit_d and $\mathsf{M}_d^{\mathrm{BiCrit}}$, see Milnor's article [79]. At this time, it does not appear to be known whether the semistable completion of $\mathsf{M}_d^{\mathrm{BiCrit}}$ coming from geometric invariant theory is isomorphic to the projective closure of the variety (2.13).

[4]To avoid problems with wild ramification, we restrict attention to fields of characteristic p with either $p = 0$ or $p > d$.

CHAPTER 3

Dynamical Moduli Spaces—Further Topics

3.1. An application to good reduction over function fields

In this section we give an application of the existence of the moduli space M_d^n to the study of isotriviality of families of self-maps of $\mathbb{P}^n$. Let k be an algebraically closed field and let $K = k(C)$ be the function field of a smooth projective curve C/k. The set of absolute values M_K on K corresponds to the set points $t \in C(k)$ via

$$|a|_t = e^{-\operatorname{ord}_t(a)} \quad \text{for } a \in K,$$

where we view a as a function on C, and where $\operatorname{ord}_t(a)$ is the order of vanishing of the function a at the point t.

A morphism $\phi \in \operatorname{Hom}_d^n(K)$ may be spread out to obtain a family of maps parametrized by the points of C,

$$\mathbb{P}_C^n \overset{\Phi}{\dashrightarrow} \mathbb{P}_C^n,$$

where $\mathbb{P}_C^n = \mathbb{P}^n \times C$. We note that in general, Φ is only a rational self-map of $\mathbb{P}_C^n = \mathbb{P}^n \times C$, not a morphism.

We define three important properties that ϕ may possess.

Definition 3.1. A map $\phi \in \operatorname{Hom}_d^n(K)$ has *good reduction* at $t \in M_K$ if there is a neighborhood $U \subset C$ of t and a change of variables $f \in \mathrm{PGL}_{n+1}(K)$ such that the map $\phi^f \in \operatorname{Hom}_d^n(K)$ extends to a morphism

$$\mathbb{P}_U^n \xrightarrow{\Phi_U} \mathbb{P}_U^n.$$

We say that ϕ has *everywhere good reduction* if it has good reduction at every $t \in M_K$.

The map ϕ is said to have *everywhere potential good reduction* if there is a finite extension K'/K such that ϕ has everywhere good reduction when considered as a map defined over K.

Definition 3.2. A map $\phi \in \operatorname{Hom}_d^n(K)$ is *trivial* if there is a change of variables $f \in \mathrm{PGL}_{n+1}(K)$ and a map $\phi_0 \in \operatorname{Hom}_d^n(k)$ defined over k such that ϕ^f is the extension of ϕ_0 from k to K. The map ϕ is *isotrivial* if ϕ is trivial over some finite extension K'/K.

Definition 3.3. A map $\phi \in \operatorname{Hom}_d^n(K)$ has *constant moduli* if the induced map

$$C(k) \dashrightarrow \mathsf{M}_d^n(k), \quad t \longmapsto \langle \phi_t \rangle,$$

is constant.

Theorem 3.4 (Petsche, Szpiro, Tepper [95])). *Let $K = k(C)$ be the function field of a smooth projective curve over an algebraically closed field, and let $\phi \in \operatorname{Hom}_d^n(K)$. Then the following are equivalent:*

(a) *ϕ has everywhere potential good reduction.*
(b) *ϕ has constant moduli.*
(c) *ϕ is isotrivial.*

Remark 3.5. Benedetto [13] and Baker [2] originally proved Theorem 3.4 in dimension one using non-Archimedean analysis. In [95], Petsche, Szpiro, and Tepper give two proofs of Theorem 3.4, one of which we sketch, the other of which generalizes the methods of Benedetto and Baker.

PROOF SKETCH. **(a) ⇒ (b)** We suppose that ϕ has everywhere potential good reduction. Replacing K by a finite extension, we may assume that ϕ has everywhere good reduction over K. This means that we can cover $C(k)$ by a finite collection of open sets $U_1, \dots, U_r$ and find maps $f_1, \dots, f_r \in \mathrm{PGL}_{n+1}(K)$ such that each ϕ^{f_i} extends to a morphism on $\mathbb{P}^n_{U_i}$. In other words, we get morphisms

$$U_i \longrightarrow \mathrm{Hom}^n_d(k), \quad t \longrightarrow (\phi^{f_i})_t.$$

Composing these with the projection map $\mathrm{Hom}^n_d \to \mathsf{M}^n_d$ gives morphisms

$$U_i \longrightarrow \mathsf{M}^n_d(k), \quad t \longrightarrow \big\langle (\phi^{f_i})_t \big\rangle.$$

However, conjugating ϕ does not change the corresponding point in the moduli space. Thus on the overlaps we have

$$\big\langle (\phi^{f_i})_t \big\rangle = \big\langle (\phi^{f_j})_t \big\rangle \quad \text{for all } t \in U_i \cap U_j.$$

This shows that the maps on the U_i fit together to give a *morphism*

$$C \longrightarrow \mathsf{M}^n_d.$$

But C is projective and M^n_d is affine, so this map must be a constant, which completes the proof that ϕ has constant moduli.

(b) ⇒ (c) We assume that ϕ has constant moduli, i.e.,

$$\langle \phi \rangle \in \mathsf{M}^n_d(k) \subset \mathsf{M}^n_d(K).$$

Let $\psi \in \mathrm{Hom}^n_d(k)$ be a map defined over k such that

$$\langle \phi \rangle = \langle \psi_K \rangle,$$

where ψ_K denotes the base extension of ψ to K.

The moduli space M^n_d is a geometric quotient, so the fibers of

$$\mathrm{Hom}^n_d(\overline{K}) \longrightarrow \mathsf{M}^n_d(\overline{K})$$

consist of closed $\mathrm{PGL}_{n+1}(\overline{K})$-conjugacy classes. Hence ϕ is $\mathrm{PGL}_{n+1}(\overline{K})$ conjugate to the map ψ_K defined over k. This conjugacy takes place over some finite extension K'/K. Hence ϕ is trivial over K', so it is isotrivial over K.

(c) ⇒ (a) Finally, we suppose that ϕ is isotrivial, so after extending K and replacing ϕ by a conjugate, we may assume that $\phi \in \mathrm{Hom}^n_d(k)$. Then the map $\phi : \mathbb{P}^1_C \to \mathbb{P}^1_C$ is simply

$$\mathbb{P}^1_k \times C \longrightarrow \mathbb{P}^1_k \times C, \quad (x,t) \longrightarrow (\phi(x), t),$$

which is clearly a morphism. Hence ϕ has everyone good reduction. □

Remark 3.6. In the theory of abelian varieties, the criterion of Néron–Ogg–Shafarevich [101] says that good reduction is equivalent to the inertia subgroup of the Galois group acting trivially on points of finite order. It would be interesting to formulate a dynamical analogue of this criterion, say in terms of the action of inertia on periodic or preperiodic points.

3.2. Minimal resultants and minimal models

In Section 3.1 we defined what it means for a map $\phi \in \operatorname{Hom}_d^n(K)$ defined over a function field K to have good reduction at a place of K. This definition makes sense scheme-theoretically for the field of fractions of any Dedekind domain. We start with the local case.

Definition 3.7. Let R be a discrete valuation ring with field of fractions K. A map $\phi \in \operatorname{Hom}_d^n(K)$ induces a rational map

$$\phi_R \colon \mathbb{P}_R^n \dashrightarrow \mathbb{P}_R^n$$

over $\operatorname{Spec} R$. We say that ϕ has *good reduction* if there is a change of variables $f \in \operatorname{PGL}_{n+1}(K)$ such that ϕ_R^f extends to a morphism. We say that ϕ has *potential good reduction* if there is a finite extension R'/R with field of fractions K' such that ϕ has good reduction as an element of $\operatorname{Hom}_d^n(K')$.

The construction of the moduli space M_d^n as a scheme over $\operatorname{Spec}\mathbb{Z}$ leads to the following characterization of potential good reduction.

Proposition 3.8. *Let R be a discrete valuation ring with field of fractions K, and let $\phi \in \operatorname{Hom}_d^n(K)$. Then*

$$\phi \textit{ has potential good reduction} \iff \langle\phi\rangle \in \mathsf{M}_d^n(R).$$

PROOF. Suppose first that ϕ has good reduction, so there is an $f \in \operatorname{PGL}_2(K)$ such that ϕ_R^f extends to a self-morphism of $\mathbb{P}_R^n$. This means that $\phi^f \in \operatorname{Hom}_d^n(R)$, so

$$\langle\phi\rangle = \langle\phi^f\rangle \in \mathsf{M}_d^n(R).$$

More generally, if ϕ only has potential good reduction, then the same argument says that there is a finite extension R'/R such that $\langle\phi\rangle \in \mathsf{M}_d^n(R')$. But ϕ is defined over K, so we also have $\langle\phi\rangle \in \mathsf{M}_d^n(K)$. Since $R' \cap K = R$, this proves that $\langle\phi\rangle \in \mathsf{M}_d^n(R)$.

Conversely, suppose that $\langle\phi\rangle \in \mathsf{M}_d^n(R)$. This means that there is some finite extension K'/K and some $f \in \operatorname{PGL}_2(K')$ such that $\phi^f \in \operatorname{Hom}_d^n(R')$, where R' is the ring of integers of K'. Then

$$\phi_{R'}^f \colon \mathbb{P}_{R'}^n \longrightarrow \mathbb{P}_{R'}^n$$

is a morphism over R', which proves that ϕ has potential good reduction. □

We can use the theory of resultants to give a nice characterization of good reduction. We recall from Theorem 1.8 that the Macaulay resultant $\mathcal{R}$ of a set of homogeneous polynomials $\phi_0, \ldots, \phi_n \in K[X_0, \ldots, X_n]$ of degree d is a polynomial in the coefficients of $\phi_0, \ldots, \phi_n$ whose vanishing is equivalent to $\phi_0, \ldots, \phi_n$ having a common nontrivial zero in $\mathbb{A}^{n+1}(\overline{K})$.

Definition 3.9. Let R be a discrete valuation ring with valuation v and field of fractions K. For any $\phi = [\phi_0, \dots, \phi_n] \in \mathrm{Hom}_d^n(K)$, we define

$$e_v(\phi) = v\big(\mathcal{R}(\phi)\big) - (n+1)d^n \min_{0 \le i \le n} v(\phi_i),$$

where $v(\phi_i)$ denotes the minimum of the valuations of the coefficients of ϕ_i. Then the *exponent at v of the minimal resultant of ϕ* is defined to be the quantity

$$\varepsilon_v(\phi) = \min_{f \in \mathrm{PGL}_2(K)} e_v(\phi^f).$$

Remark 3.10. The quantity $e_v(\phi)$ does not depend on the lift of ϕ to a self-map of $\mathbb{A}^{n+1}$, because the Macaulay resultant is multihomogeneous of degree d^n in the coefficients of each of the polynomials $\phi_0, \dots, \phi_n$; cf. Remark 1.9. For a general homogeneity formula in which $\phi_0, \dots, \phi_n$ may have different degrees, see [25, page 25, first displayed formula].

Proposition 3.11. *Let R be a discrete valuation ring with valuation v and field of fractions K, and let $\phi \in \mathrm{Hom}_d^n(K)$. Then*

$$\phi \textit{ has good reduction} \iff \varepsilon_v(\phi) = 0.$$

PROOF. Suppose first that ϕ has good reduction. Replacing ϕ by a conjugate, we may assume that ϕ extends to a morphism $\phi_R : \mathbb{P}_R^n \to \mathbb{P}_R^n$ over the ring R. This means that we can write ϕ as

$$\phi_R = [\phi_0, \dots, \phi_n] \quad \text{with } \phi_0, \dots, \phi_n \in R[X_0, \dots, X_n]$$

so that ϕ remains a morphism when restricted to the special fiber. Writing $\mathfrak{p}$ for the maximal ideal of R and $k = R/\mathfrak{p}$ for the residue field, this means that if we reduce the coefficients of the ϕ_i modulo $\mathfrak{p}$, we obtain a *morphism*

$$\tilde{\phi}_R = [\tilde{\phi}_0, \dots, \tilde{\phi}_n] \colon \mathbb{P}_k^n \longrightarrow \mathbb{P}_k^n.$$

This implies first that $\min v(\phi_i) = 0$, and second that the polynomials $\tilde{\phi}_0, \dots, \tilde{\phi}_n$ have no common zeros in $k^n \smallsetminus \{0\}$. Therefore the Macaulay resultant satisfies

$$\mathcal{R}(\tilde{\phi}_0, \dots, \tilde{\phi}_n) \ne 0 \quad \text{in } k.$$

Since

$$\mathcal{R}(\tilde{\phi}_0, \dots, \tilde{\phi}_n) = \widetilde{\mathcal{R}(\phi_0, \dots, \phi_n)},$$

it follows that $v\big(\mathcal{R}(\phi_0, \dots, \phi_n)\big) = 0$. Hence

$$\varepsilon_v(\phi) = e_v(\phi) = v\big(\mathcal{R}(\phi)\big) - (n+1)d^n \min_{0 \le i \le n} v(\phi_i) = 0.$$

Conversely, suppose that $\varepsilon_v(\phi) = 0$. Replacing ϕ by a conjugate, we may assume that $e_v(\phi) = 0$. Next, replacing ϕ by $c\phi$, we may assume that all of the ϕ_i have coefficients in R and that one of the coefficients is a unit, i.e., we may assume that $\min v(\phi_i) = 0$. We are then in the situation that

$$0 = \varepsilon_v(\phi) = e_v(\phi) = v\big(\mathcal{R}(\phi)\big),$$

so the Macaulay resultant $\mathcal{R}(\phi)$ is a unit in R. It follows that when we reduce ϕ modulo $\mathfrak{p}$,

$$[\tilde{\phi}_0, \dots, \tilde{\phi}_n] \colon \mathbb{P}_k^n \longrightarrow \mathbb{P}_k^n,$$

we get a morphism. Therefore $\phi_R \colon \mathbb{P}_R^1 \to \mathbb{P}_R^1$ is a morphism, and hence ϕ has good reduction. $\square$

Definition 3.12. Let R be a Dedekind domain and let K be its field of fractions. We say that a map $\phi \in \mathrm{Hom}_d^n(K)$ has *everywhere good reduction* if it has good reduction over every localization of R, and similarly with *potential good reduction.* The *minimal resultant of* ϕ is the ideal

$$\mathfrak{R}(\phi) = \prod_v \mathfrak{p}_v^{\varepsilon_v(\phi)},$$

where v runs over the inequivalent discrete valuations on R, and $\mathfrak{p}_v$ is the prime ideal of R associated to v. Proposition 3.11 implies that ϕ has everywhere good reduction if and only if $\mathfrak{R}(\phi) = (1)$.

For elliptic curves there is a nice characterization of everywhere good reduction in terms of the triviality of a certain ideal class. (See [111, VIII.8.2].) There is an analogous result for rational functions on $\mathbb{P}^1$, although it is only in one direction. Presumably a similar result holds in higher dimension.

Proposition 3.13. *Let R be a Dedekind domain with field of fractions K, let $\phi \in \mathrm{Hom}_d^1(K)$, and write $\phi = [F, G]$ with homogeneous polyomials $F, G, \in K[X, Y]$.*

(a) *There is a fractional ideal $\mathfrak{a}(F, G)$ of K satisfying*

$$\mathfrak{R}(\phi) = \begin{cases} \mathrm{Resultant}(F,G)\mathfrak{a}(F,G)^{2d} & \text{if } d \text{ is odd,} \\ \mathrm{Resultant}(F,G)\mathfrak{a}(F,G)^{d} & \text{if } d \text{ is even.} \end{cases}$$

(b) *Let $n(d) = 1$ (respectively $n(d) = 2$) if d is odd (respectively even). Let I_K be the group of fractional ideals of K, and let P_K be the group of principal ideals. Then the image of $\mathfrak{a}(F, G)$ in $I_K/P_K^{n(d)}$ depends only on ϕ, and not on the choice of F and G. We denote this image by $\bar{\mathfrak{a}}(\phi)$.*

(c) *If ϕ has a global minimal model over K, then $\bar{\mathfrak{a}}(\phi) = (1)$.*

PROOF. See [110, Propositions 4.99 and 4.100]. □

Question 3.14. We pose some questions related to minimal resultants and good reduction. As usual, we assume that all maps have degree at least 2.

(a) Is the converse to Proposition 3.13 true, i.e., if $\bar{\mathfrak{a}}(\phi) = (1)$, does ϕ necessarily have a global minimal model over K?

(b) As a special case of (a), is it true that every rational map $\phi(z) \in \mathbb{Q}(z)$ has a global minimal model over $\mathbb{Q}$? If not, how about maps of odd degree? of even degree?

(c) This question relates to the $\overline{K}/K$-twists of a rational map; see Definition 7.3 in Section 7.1 for the definition of the set of twists. Let K be a number field and let $\phi(z) \in K(z)$. Let S be a finite set of primes of K. Is the set

$$\{\psi \in \mathrm{Twist}_K(\phi) : \psi \text{ has good reduction at all } \mathfrak{p} \notin S\}$$

finite?

Remark 3.15. A theorem of Shafarevich [111, IX.6.1] says that there are only finitely many elliptic curves defined over a given number field K having good reduction outside of a given finite set of primes S, and more generally, Faltings [41] proved that the same is true for abelian varieties of a given dimension. There is no obvious dynamical analogue to these results, since for example, every monic polynomial in $\mathbb{Z}[z]$ has everywhere good reduction. See [116] for a related (dynamical) result for rational map $\phi \in \mathrm{Hom}_d$ under the equivalence relation $\phi \sim g \circ \phi \circ f$ for $f, g\,\mathrm{PGL}_2$.

3.3. Dynamics on K3 Surfaces

Projective spaces admit many self-maps, so they are interesting spaces on which to study dynamics. But there are many other varieties that have self-maps of infinite order. In this section we briefly discuss K3 surfaces possessing two (or more) noncommuting involutions ι_1 and ι_2 whose composition $\iota_1 \circ \iota_2$ defines an automorphism of infinite order. For further reading on the material in this section, see for example [75, 105, 110]; and see Section 5.8 for a discussion of canonical heights on K3 surfaces.

In general, suppose that V is a variety with an involution

$$\iota : V \longrightarrow V,$$

i.e., an automorphism whose second iterate is the identity,

$$\iota^2(x) = x \quad \text{for all } x \in V.$$

If we take the quotient of V by the relation

$$\iota(x) \sim x,$$

we obtain a quotient variety $W = V/\iota$ and a double cover

$$\pi : V \longrightarrow W.$$

The involution ι is determined by the condition

$$\pi^{-1}\bigl(\pi(x)\bigr) = \bigl\{x, \iota(x)\bigr\}.$$

Conversely, a double cover $V \to W$ induces an involution defined by switching the sheets of the covering.

The dynamics of a single involution is not very interesting, but some varieties have noncommuting involutions ι_1 and ι_2 whose composition

$$\phi = \iota_1 \circ \iota_2$$

is an automorphism of infinite order. In such situations, the dynamics of ϕ can be extremely interesting.

We will look at a specific family of varieties of this sort. We start with two bihomogeneous polynomials, one of bidegree $(1,1)$ and the other of bidegree $(2,2)$,

$$L(\boldsymbol{x}, \boldsymbol{y}) = \sum_{i=0}^{2}\sum_{j=0}^{2} A_{ij}x_i y_j, \quad Q(\boldsymbol{x}, \boldsymbol{y}) = \sum_{0\le i\le j\le 2}\;\sum_{0\le k\le \ell\le 2} B_{ijk\ell}x_i x_j y_k y_\ell,$$

and we take $S \subset \mathbb{P}^2 \times \mathbb{P}^2$ to be the surface defined by these polynomials. The surface S is determined by the 45 coefficients of L and Q,

$$\boldsymbol{A} = [A_{00}, A_{01}, \dots, A_{22}] \in \mathbb{P}^8 \quad \text{and} \quad \boldsymbol{B} = [B_{0000}, B_{0001}, \dots, B_{2222}] \in \mathbb{P}^{35},$$

so we write

$$S_{\boldsymbol{A},\boldsymbol{B}} = \bigl\{(\boldsymbol{x}, \boldsymbol{y}) \in \mathbb{P}^2 \times \mathbb{P}^2 : L_{\boldsymbol{A}}(\boldsymbol{x}, \boldsymbol{y}) = Q_{\boldsymbol{B}}(\boldsymbol{x}, \boldsymbol{y}) = 0\bigr\}$$

to indicate the dependence on $\boldsymbol{A}$ and $\boldsymbol{B}$. The parameters $\boldsymbol{A}$ and $\boldsymbol{B}$ that determine $S_{\boldsymbol{A},\boldsymbol{B}}$ live in

$$(\boldsymbol{A}, \boldsymbol{B}) \in \mathbb{P}^8 \times \mathbb{P}^{35}.$$

The embedding of $S_{\boldsymbol{A},\boldsymbol{B}}$ in $\mathbb{P}^2 \times \mathbb{P}^2$ gives two projection maps

$$\pi_1, \pi_2 \colon S \longrightarrow \mathbb{P}^2, \qquad \pi_1(\boldsymbol{x}, \boldsymbol{y}) = \boldsymbol{x}, \quad \pi_2(\boldsymbol{x}, \boldsymbol{y}) = \boldsymbol{y}.$$

These projections are (generically) of degree two, so as described above they induce (rational) involutions

$$\iota_1, \iota_2 \colon S_{\boldsymbol{A},\boldsymbol{B}} \longrightarrow S_{\boldsymbol{A},\boldsymbol{B}}.$$

More precisely, the involution ι_1 is well-defined at $x \in S_{\boldsymbol{A},\boldsymbol{B}}$ provided that the fiber $\pi_1^{-1}\big(\pi_1(x)\big)$ is zero-dimensional (and hence consists of one or two points), and similarly for ι_2.

Proposition 3.16. *There is a Zariski open subset $U \subset \mathbb{P}^8 \times \mathbb{P}^{35}$ with the property that for every $\boldsymbol{A}, \boldsymbol{B} \in U$, the surface $S_{\boldsymbol{A},\boldsymbol{B}}$ is nonsingular and the involutions $\iota_1, \iota_2 \in \operatorname{Aut}(S_{\boldsymbol{A},\boldsymbol{B}})$ are morphisms, i.e., are well-defined on all of $S_{\boldsymbol{A},\boldsymbol{B}}$.*

PROOF. See [110, §7.4.1] for a computational proof. In fact, the nonsingularity is enough to ensure that ι_1 and ι_2 are morphisms; see Proposition 3.18. □

The involutions ι_1 and ι_2 do not commute, and their composition

$$\phi = \iota_1 \circ \iota_2 \in \operatorname{Aut}(S_{\boldsymbol{A},\boldsymbol{B}})$$

is an automorphism of $S_{\boldsymbol{A},\boldsymbol{B}}$ of infinite order. One can prove that ι_1 and ι_2 satisfy no relations other than $\iota_1^2 = \iota_2^2 = 1$. We will discuss some of the dynamics of ϕ in Section 5.8, but for now we are interested in the moduli space of pairs $(S_{\boldsymbol{A},\boldsymbol{B}}, \phi)$.

The smooth surfaces $S_{\boldsymbol{A},\boldsymbol{B}}$ on which ϕ is regular are parametrized by the points in an open subset U of $\mathbb{P}^8 \times \mathbb{P}^{35}$, but many of these surfaces are isomorphic. For example, we can use

$$\boldsymbol{f} = (f_1, f_2) \in \operatorname{PGL}_3 \times \operatorname{PGL}_3$$

to change variables on the two factors of $\mathbb{P}^2$. Then $S_{\boldsymbol{A},\boldsymbol{B}}$ is isomorphic to the surface given by the equations

$$S_{\boldsymbol{A}^{\boldsymbol{f}},\boldsymbol{B}^{\boldsymbol{f}}} : L_{\boldsymbol{A}}\big(f_1^{-1}(\boldsymbol{x}), f_2^{-1}(\boldsymbol{y})\big) = Q_{\boldsymbol{B}}\big(f_1^{-1}(\boldsymbol{x}), f_2^{-1}(\boldsymbol{y})\big) = 0.$$

The involutions on $S_{\boldsymbol{A}^{\boldsymbol{f}},\boldsymbol{B}^{\boldsymbol{f}}}$ are then given by the usual conjugation formulas,

$$\iota_1^{\boldsymbol{f}}(\boldsymbol{x}, \boldsymbol{y}) = \boldsymbol{f}^{-1} \circ \iota_1 \circ \boldsymbol{f}(\boldsymbol{x}, \boldsymbol{y}) \quad \text{and} \quad \iota_2^{\boldsymbol{f}}(\boldsymbol{x}, \boldsymbol{y}) = \boldsymbol{f}^{-1} \circ \iota_2 \circ \boldsymbol{f}(\boldsymbol{x}, \boldsymbol{y}),$$

and the automorphism of infinite order on $S_{\boldsymbol{A}^{\boldsymbol{f}},\boldsymbol{B}^{\boldsymbol{f}}}$ is

$$\phi^{\boldsymbol{f}} = \iota_1^{\boldsymbol{f}} \circ \iota_2^{\boldsymbol{f}} = \boldsymbol{f}^{-1} \circ \iota_1 \circ \iota_2 \circ \boldsymbol{f}.$$

The parameter space $\mathbb{P}^8 \times \mathbb{P}^{35}$ has dimension 43, and each copy of PGL_3 has dimension 8, so we expect the quotient to have dimension 27. However, we have not exhausted all of the ways in which different values of the parameters $(\boldsymbol{A}, \boldsymbol{B})$ can give isomorphic surfaces. Roughly speaking, we can add a multiple of $\boldsymbol{A}$ to $\boldsymbol{B}$ and get the same surface, since the surface $S_{\boldsymbol{A},\boldsymbol{B}}$ is determined by the homogeneous ideal generated by $L_{\boldsymbol{A}}$ and $Q_{\boldsymbol{B}}$. However, we need to be a bit careful, because $\boldsymbol{A}$ and $\boldsymbol{B}$ are points in projective space, so they only determine $L_{\boldsymbol{A}}$ and $Q_{\boldsymbol{B}}$ up to multiplication by a scalar.

Thus for $(\boldsymbol{A}, \boldsymbol{B}) \in \mathbb{P}^8 \times \mathbb{P}^{35}$, we choose points

$$\mathbf{a} \in \mathbb{A}^9 \smallsetminus \{0\} \quad \text{and} \quad \mathbf{b} \in \mathbb{A}^{36} \smallsetminus \{0\}$$

that map to $\boldsymbol{A}$ and $\boldsymbol{B}$. Then $L_{\mathbf{a}}$ and $Q_{\mathbf{b}}$ are specific forms. Further, any bihomogeneous $(1,1)$ form $M(\boldsymbol{x}, \boldsymbol{y})$ gives a surface defined by the equations

$$L_{\mathbf{a}}(\boldsymbol{x}, \boldsymbol{y}) = 0 \quad \text{and} \quad Q_{\mathbf{b}}(\boldsymbol{x}, \boldsymbol{y}) + M(\boldsymbol{x}, \boldsymbol{y}) L_{\mathbf{a}}(\boldsymbol{x}, \boldsymbol{y}) = 0,$$

that is is identical to the surface $S_{\boldsymbol{A},\boldsymbol{B}}$. This is because the surface $S_{\boldsymbol{A},\boldsymbol{B}}$ is determined by the homogeneous ideal generated by $L_{\mathbf{a}}$ and $Q_{\mathbf{b}}$, which is clearly the same as the ideal generated by $L_{\mathbf{a}}$ and $Q_{\mathbf{b}} + ML_{\mathbf{a}}$.

The space of $(1,1)$ forms M has dimension 9, so the moduli space of isomorphism classes of surfaces $(S_{\boldsymbol{A},\boldsymbol{B}},\phi)$ has dimension at most

$$\underbrace{8}_{\boldsymbol{A}\in\mathbb{P}^8} + \underbrace{35}_{\boldsymbol{B}\in\mathbb{P}^{35}} - \underbrace{8}_{\mathrm{PGL}_3} - \underbrace{8}_{\mathrm{PGL}_3} - \underbrace{9}_{M} = 18.$$

One can prove that these are the only isomorphisms between the various $S_{\boldsymbol{A},\boldsymbol{B}}$, so there is an 18-parameter family of isomorphism classes of nonsingular surfaces $S_{\boldsymbol{A},\boldsymbol{B}}$.

Remark 3.17. The surface $S_{\boldsymbol{A},\boldsymbol{B}}$, assuming that it is nonsingular, is an example of a K3 surface. These are regular surfaces with trivial canonical bundle, where we recall that a surface S is regular if $h^1(S,\mathcal{O}_S)=0$. The moduli space of complex (not necessarily projective) K3 surfaces is 20-dimensional, and within that space, the set of projective K3 surfaces forms a countable family of 19-dimensional subvarieties. More precisely, for each $g\ge 2$ there is a 19-dimensional moduli space F_g classifying isomorphism classes of K3 surfaces that have a primitive ample line bundle $\mathcal{L}(C)$, where C is a curve of genus g. It is known that F_g is unirational for $g\le 13$ and of general type for $g\ge 63$. See [119] for a survey of what is known about the geometry of F_g.

Restricting to the case $g=2$, every K3 surface classified by F_2 can be realized as a double cover of $\mathbb{P}^2$ branched along a nonsingular curve of degree 6. The nonsingular surfaces $S_{\boldsymbol{A},\boldsymbol{B}}$ have this property, but they are special because they are actually double covers of $\mathbb{P}^2$ in two different ways. In this way, the set of isomorphism classes of surfaces $S_{\boldsymbol{A},\boldsymbol{B}}$ form an irreducible 18-dimensional subvariety of F_2.

The next proposition shows that nonsingularity of $S_{\boldsymbol{A},\boldsymbol{B}}$ suffices to ensure that the involutions ι_1 and ι_2 are morphisms.

Proposition 3.18. *Let S and S' be nonsingular projective* K3 *surfaces and let $\phi\colon S' \dashrightarrow S$ be a birational map. Then ϕ is an automorphism.*

PROOF SKETCH. Let U' be the largest open subset of S' on which ϕ is defined, so in particular $S'\smallsetminus U'$ consists of a finite set of points. The surface S has trivial canonical bundle, so we can find a a nonvanishing holomorphic 2-form ω_S on S. Then

$$\phi^*\omega_S$$

is a 2-form on S' that might have zeros and/or poles, but in any case it is holomorphic on U'. However, the poles of a 2-form on S' are divisors, i.e., have codimension 1, so $\phi^*\omega_S$ extends to a holomorphic 2-form $\omega_{S'}$ on S'. Since S' also has trivial canonical bundle, it follows that $\omega_{S'}$ is a nonvanishing holomorphic 2-form on S'.

Suppose that there is an exceptional curve $E'\subset S'$, i.e., such that $\phi(E')$ consists of a single point. Then $\phi^*\omega_S$ would vanish on E'. This contradicts the fact that $\phi^*\omega_S$ coincides with the nonvanishing form $\omega_{S'}$ on the set U', whose complement consists of a finite set of points. Therefore ϕ does not contract any curves down to a point.

Suppose now that ϕ^{-1} is not a morphism. Then there is a point $P\in S$ and a curve $E'\subset S'$ with $\phi(S')=P$, which is a contradiction. This proves that ϕ^{-1}

is a morphism, and repeating the argument with ϕ and ϕ^{-1} interchanged gives the same result for ϕ. $\square$

Example 3.19. There are many other ways to construct K3 surfaces with interesting automorphisms. For example, let

$$S \subset \mathbb{P}^1 \times \mathbb{P}^1 \times \mathbb{P}^1$$

be a hypersurface given by the vanishing of a form $L(\boldsymbol{x}, \boldsymbol{y}, \boldsymbol{z})$ of type $(2, 2, 2)$, i.e.,

$$L([x_0, x_1], [y_0, y_1], [z_0, z_1]) = \sum C_{ijk} x_0^i x_1^{2-i} y_0^j y_1^{2-j} z_0^k z_1^{2-k}.$$

The three projections

$$p_{12}, p_{13}, p_{23} \colon \mathbb{P}^1 \times \mathbb{P}^1 \times \mathbb{P}^1 \longrightarrow \mathbb{P}^1 \times \mathbb{P}^1,$$
$$p_{12}(\boldsymbol{x}, \boldsymbol{y}, \boldsymbol{z}) = (\boldsymbol{x}, \boldsymbol{y}), \quad p_{13}(\boldsymbol{x}, \boldsymbol{y}, \boldsymbol{z}) = (\boldsymbol{x}, \boldsymbol{z}), \quad p_{23}(\boldsymbol{x}, \boldsymbol{y}, \boldsymbol{z}) = (\boldsymbol{y}, \boldsymbol{z}),$$

induce double covers $S \to \mathbb{P}^1 \times \mathbb{P}^1$, which in turn induce involutions

$$\iota_{12}, \iota_{13}, \iota_{23} \colon S \longrightarrow S.$$

These involutions do not satisfy any relations other than $\iota_{jk}^2 = 1$, so compositions such as $\phi = \iota_{12} \circ \iota_{13}$ and $\psi = \iota_{12} \circ \iota_{23}$ are noncommuting automorphisms of infinite order. The dynamics of each individual map such as ϕ or ψ is interesting, and it is also interesting to look at the full orbit of a point under the subgroup of $\operatorname{Aut}(S)$ generated by all three involutions ι_1, ι_2, and ι_3. See [5, 6, 8, 121] for material related to dynamics on these K3 surfaces.

3.4. An algebraic characterization of Lattès maps

The set of flexible Lattès maps appears as an exceptional family in many dynamical situations. For example, McMullen's theorem (Theorem 2.38) says that the flexible Lattès maps are the only infinite family of maps in M_d having identical multiplier systems, and thus define the only curves in M_d that are compressed to a point under the Milnor map $\boldsymbol{s}_N : \mathsf{M}_d(\mathbb{C}) \to \mathbb{C}^L$. We recall the general definition of a Lattès map.

Definition 3.20. A rational map $\phi \colon \mathbb{P}^1 \to \mathbb{P}^1$ is a *Lattès map* if there is an elliptic curve E, an endomorphism $\psi \colon E \to E$, and a finite map $\pi \colon E \to \mathbb{P}^1$ such that the following diagram commutes:

$$\begin{array}{ccc} E & \xrightarrow{\psi} & E \\ \downarrow{\scriptstyle \pi} & & \downarrow{\scriptstyle \pi} \\ \mathbb{P}^1 & \xrightarrow{\phi} & \mathbb{P}^1. \end{array}$$

We say that the triple (E, ψ, π) is a *realization of* ϕ. A Lattès map is *flexible* if it varies in a continuous family, otherwise it is *rigid*.

Remark 3.21. Flexible Lattès maps are associated to multiplication-by-m maps on elliptic curves, and thus their degrees are squares. Hence if d is not a square, then Hom_d and M_d have no points associated to flexible Lattès maps.

The following theorem (and considerably more) was proven by Milnor [80] over $\mathbb{C}$ using analytic and geometric ideas, and by Ghioca and Zieve [44] in arbitrary characteristic using algebraic methods. We sketch the latter proof.

Theorem 3.22 (Ghioca–Zieve [44]). *Let $\phi\colon \mathbb{P}^1 \to \mathbb{P}^1$ be a Lattès map of degree $d \ge 5$. Then there is a realization (E, ψ, π) for ϕ with the property that the map $\pi\colon E \to \mathbb{P}^1$ is Galois. In particular, there is a nontrivial subgroup $\Gamma \subset E$ such that π factors as*

$$E \longrightarrow E/\Gamma \cong \mathbb{P}^1.$$

If $j(E) \ne 0$ and $j(E) \ne 1728$, then necessarily $\Gamma = \{\pm 1\}$.

We begin with two preliminary results.

Lemma 3.23. *Let E be a curve of genus 1 and let $\psi\colon E \to \mathbb{P}^1$ be a finite separable isogeny, so $K(E)/\psi^* K(\mathbb{P}^1)$ is a finite separable field extension. Then the field*

$$L' = \bigcap_{\substack{\psi^* K(\mathbb{P}^1) \subset L \subset K(E) \\ \text{genus}(L)=1}} L \tag{3.1}$$

is a field of genus 1.

Proof. Without loss of generality, we may assume that the constant field is algebraically closed, so curves of genus 1 can be given the structure of elliptic curves. Let L be one of the fields in the intersection (3.1), so $L = K(C_L)$ for some elliptic curve C_L and there are finite maps giving a commutative diagram

$$E \xrightarrow{\alpha_L} C_L \xrightarrow{\beta_L} \mathbb{P}^1, \qquad \psi = \beta_L \circ \alpha_L.$$

The map $\alpha_L\colon E \to C_L$ is Galois, since any separable map between elliptic curves is Galois, and further $\operatorname{Gal}\bigl(K(E)/\alpha_L^* K(C_L)\bigr) \cong \ker(\alpha_L)$; see [111, III.4.10]. Let

$$\Gamma = \sum_{\substack{\psi^* K(\mathbb{P}^1) \subset L \subset K(E) \\ \text{genus}(L)=1}} \ker(\alpha_L) \subset E$$

be the subgroup of E generated by the kernels of all of the α_L. Then Γ is a finite subgroup of E, and L' is the function field of the elliptic curve E/Γ, so L' has genus 1. □

Definition 3.24. A finite map $E \to \mathbb{P}^1$ from an elliptic curve to $\mathbb{P}^1$ is *reduced* if there are no nontrivial factorizations $E \to E' \to \mathbb{P}^1$, where E' is an elliptic curve. A realization (E, ψ, π) of a Lattès map is *reduced* if the map $\pi\colon E \to \mathbb{P}^1$ is reduced.

Lemma 3.25. *Every Lattès map admits a reduced realization.*

Proof Sketch. Lemma 3.23 says that every finite separable map $\nu\colon E \to \mathbb{P}^1$ admits a unique factorization $E \to E_\nu \to \mathbb{P}^1$ such that E_ν is reduced, i.e., any other factorization $E \to E' \to \mathbb{P}^1$ factors through ν as

$$E \longrightarrow E' \longrightarrow E_\nu \longrightarrow \mathbb{P}^1.$$

Let $\phi\colon \mathbb{P}^1 \to \mathbb{P}^1$ be a Lattès map and let (E, ψ, π) be any realization of ϕ. We are going to apply Lemma 3.23 several times to fill in the dotted arrows in Figure 3.1.

The outer square of Figure 3.1 is our original realization (E, ψ, π) for ϕ. The map from the upper left to the lower right of the square is equal to both $\phi \circ \pi$ and $\pi \circ \psi$, so that diagonal map factors through its reduced elliptic curve $E_{\phi\circ\pi} = E_{\pi\circ\psi}$.

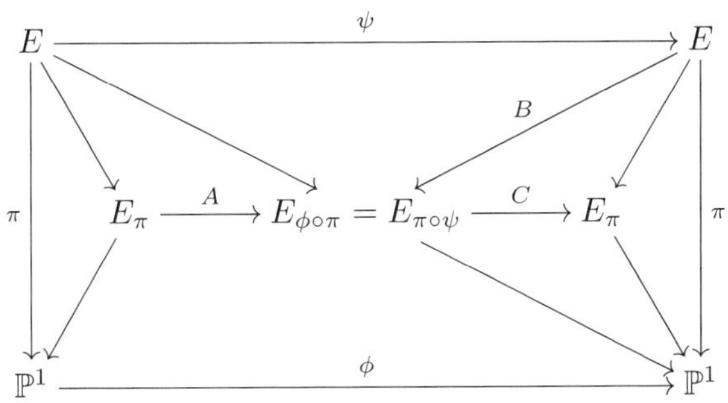

FIGURE 3.1. Constructing a reduced Lattès realization

Similarly, the vertical left and right maps are π, so they factor through their reduced elliptic curves E_π. This explains all of the arrows in Figure 3.1 except for the arrows labeled A, B, and C.

The left and bottom of the diagram give

$$E \longrightarrow E_\pi \longrightarrow \mathbb{P}^1 \xrightarrow{\ \phi\ } \mathbb{P}^1, \qquad (E \to \mathbb{P}^1 \text{ being } \pi).$$

This shows that E_π factors $\phi \circ \pi$, so by the reducedness of $E_{\phi\circ\pi}$, we get a map $E_\pi \to E_{\phi\circ\pi}$, which is labeled as A in Figure 3.1.

Next, we note that

$$E \xrightarrow{\psi} E \xrightarrow{\pi} \mathbb{P}^1$$

factors $\pi\circ\psi$, so we get the map labeled B in Figure 3.1 directly from the reducedness of $E_{\pi\circ\psi}$.

Finally, we use the fact that

$$E \xrightarrow{B} E_{\pi\circ\psi} \longrightarrow \mathbb{P}^1$$

factors π to obtain the map $C\colon E_{\pi\circ\psi} \to E_\pi$ in Figure 3.1.

We have now constructed a realization

$$\begin{array}{ccc} E_\pi & \xrightarrow{C\circ A} & E_\pi \\ \downarrow & & \downarrow \\ \mathbb{P}^1 & \xrightarrow{\ \phi\ } & \mathbb{P}^1 \end{array}$$

for ϕ, and by construction the maps $E_\pi \to \mathbb{P}^1$ are reduced, so the realization is reduced. $\square$

We are now ready to prove a strengthened version of Theorem 3.22.

Theorem 3.26. *Let ϕ be a Lattès map and let (E,ψ,π) be a reduced realization of ϕ.*

(a) *$K(E)$ contains the splitting field of $K(\mathbb{P}^1)$ over $\phi^*K(\mathbb{P}^1)$.*
(b) *If $\deg\phi \ge 5$ and the characteristic of K is not 2 or 3, then $K(E)$ is equal to the splitting field of $K(\mathbb{P}^1)/\phi^*K(\mathbb{P}^1)$.*

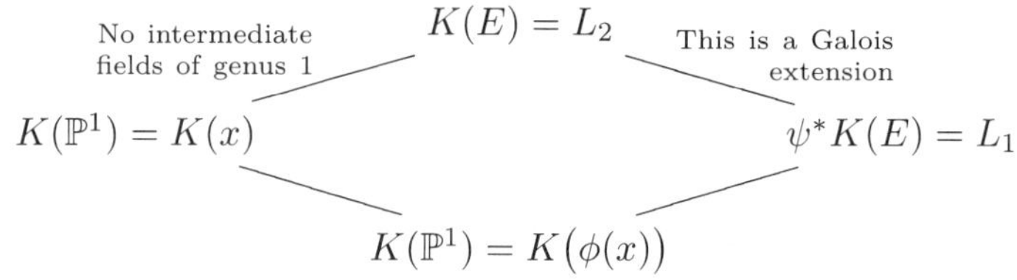

FIGURE 3.2. The fields associated to a reduced realization

PROOF SKETCH. Choosing a new origin on E, we may assume that ψ is an isogeny. To ease notation, we write

$$K(\mathbb{P}^1) = K(x), \quad \phi^* K(\mathbb{P}^1) = K\bigl(\phi(x)\bigr), \quad L_2 = K(E), \quad L_1 = \psi^* K(E).$$

This gives the diagram of fields illustrated in Figure 3.2. The extension L_2/L_1 is Galois, since it is an extension of function fields of genus one. Further, if $\tau \in \operatorname{Gal}(L_2/L_1)$, then $\tau(x) \in L_2$ is a conjugate of x over $K\bigl(\phi(x)\bigr)$, since $K\bigl(\phi(x)\bigr)$ is contained in L_2 and τ fixes L_2. We thus get a well-defined map

$$\operatorname{Gal}(L_2/L_1) \longrightarrow \left\{ \begin{matrix} K\bigl(\phi(x)\bigr) \text{ conjugates of } x \\ \text{that are contained in } L_2 \end{matrix} \right\}. \tag{3.2}$$

We claim that the map (3.2) is injective. It suffices to prove that

$$\tau(x) = x \iff \tau = 1,$$

so suppose that $\tau \in \operatorname{Gal}(L_2/L_1)$ fixes x. The Galois group $\operatorname{Gal}(L_2/L_1)$ may be identified with the kernel of the isogeny $\psi : E \to E$, so the automorphism τ is a translation of E, and the fixed field $K(E)^\tau = K\bigl(E/\langle\tau\rangle\bigr)$ is the function field of the elliptic curve $E/\langle\tau\rangle$. This gives the tower of fields

$$L_2/K(E)^\tau/K(x),$$

but the reducedness of (E, ψ, π) says that there are no nontrivial genus one fields between $K(x)$ and L_2. Therefore $K(E)^\tau = L_2$, so $\tau = 1$.

The injectivity of (3.2) implies that

$$\begin{aligned} \#\left\{ \begin{matrix} K\bigl(\phi(x)\bigr) \text{ conjugates of } x \\ \text{that are contained in } L_2 \end{matrix} \right\} &\ge \#\operatorname{Gal}(L_2/L_1) \\ &= \deg\psi = \deg\phi = \bigl[K(x) : K\bigl(\phi(x)\bigr)\bigr], \end{aligned}$$

which completes the proof that L_2 contains the splitting field of $K(x)$ over $K\bigl(\phi(x)\bigr)$.

In order to show that they are equal under suitable hypotheses, let $\Omega \subset K(E)$ be the splitting field of $K(x)$ over $K\bigl(\phi(x)\bigr)$. If $\Omega \ne K(E)$, then the reducedness of the realization implies that Ω has genus zero, so we obtain corresponding maps

$$E \longrightarrow \mathbb{P}^1 \longrightarrow \mathbb{P}^1, \qquad E \xrightarrow{\ \pi\ } \mathbb{P}^1.$$

The remainder of the proof is an intricate analysis of the very limited possibilities for the ramification of π and ϕ. See [44] for details. □

Remark 3.27. Ghioca and Zieve's paper [44] on Lattès maps contains far more than we have described. In particular, for characteristic not 2 or 3, there is a precise description of ramification conditions on ϕ that are necessary and sufficient for ϕ to be a Lattès map.

CHAPTER 4

Dynatomic Polynomials and Dynamical Modular Curves

In this chapter we restrict attention to self-maps of $\mathbb{P}^1$.

4.1. Dynatomic polynomials

Let

$$\phi\colon \mathbb{P}^1 \to \mathbb{P}^1, \quad \phi([X,Y]) = [F(X,Y), G(X,Y)],$$

where F and G are homogeneous of degree d with $\operatorname{Resultant}(F,G) \neq 0$. Iteration gives

$$\phi^n([X,Y]) = [F_n(X,Y), G_n(X,Y)] \quad \text{with } \deg F_n = \deg G_n = d^n.$$

Then

$$P \in \operatorname{Per}_n(\phi) \iff YF_n(X,Y) - XG_n(X,Y) \text{ vanishes at } P,$$

but we really want to consider only points of exact period n. We do this by mimicking the construction of the cyclotomic polynomials.

Definition 4.1. The *n-dynatomic polynomial* of ϕ is

$$\Phi_{\phi,n}(X,Y) = \prod_{k|n} \bigl(YF_k(X,Y) - XG_k(X,Y)\bigr)^{\mu(n/k)},$$

where μ is the Möbius function. We also sometimes dehomogenize and write

$$\Phi_{\phi,n}(z) = \Phi_{\phi,n}(z,1).$$

If ϕ is a polynomial, then

$$\Phi_{\phi,n}(z) = \prod_{k|n} (\phi^k(z) - z)^{\mu(n/k)}. \tag{4.1}$$

For rational functions in general, $\Phi_{\phi,n}(z)$ equals the numerator of the product (4.1) after it is written as a rational function with numerator and denominator having no common factors.

In the cyclotomic case, $z^n - 1$ has distinct roots, so it's clear that the classical cyclotomic polynomial

$$\prod_{k|n} (z^k - 1)^{\mu(n/k)}$$

is a polynomial, i.e., all of the factors with negative exponents cancel. Life is more complicated in the dynamical setting.

Example 4.2. Consider the polynomial

$$\phi(z) = z^2 - \frac{3}{4}.$$

We compute

$$\phi(z) - z = \left(z - \frac{3}{2}\right)\left(z + \frac{1}{2}\right),$$

$$\phi^2(z) - z = \left(z - \frac{3}{2}\right)\left(z + \frac{1}{2}\right)^3,$$

$$\Phi_{\phi,2}(z) = \frac{\phi^2(z) - z}{\phi(z) - z} = \left(z + \frac{1}{2}\right)^2.$$

In particular, the polynomial $\phi(z)$ has no points of exact period 2, since all of the roots of $\phi^2(z) - z$ are also roots of $\phi(z) - z$. Further, the polynomial $\phi^2(z) - z$ has a double root, as does the dynatomic polynomial $\Phi_{\phi,2}(z)$.

Definition 4.3. The roots of $\Phi_{\phi,n}(X,Y)$ are called points of *formal period* n. They include all points of exact period n, but may include points of lower period.

Theorem 4.4. *Let* $\phi \in \mathrm{Hom}_d$.

(a) $\Phi_{\phi,n}(X,Y)$ *is a polynomial for all* $n \geq 1$.

(b) *Let* $P \in \mathrm{Per}(\phi)$ *have exact period* m *and formal period* n, *and let* $\lambda_\phi(P) = (\phi^m)'(P)$ *be the multiplier at* P. *Then one of the following is true*:

(i) $n = m$.

(ii) $n = mr$ *with* $\lambda_\phi(P)$ *a primitive* r*th-root of unity.*

(iii) $n = mrp^e$ *with* r *as in* (ii) *and* $\mathrm{char}(K) = p$.

Proof ideas. First study $\mathrm{ord}_P\,\Phi_{\phi,n}$ for a fixed point P by using the Taylor series expansion. Then do a detailed case-by-case analysis replacing ϕ by ϕ^k for various values of k, using the obvious identity

$$\mathrm{ord}_P\,\Phi_{\phi,km} = \mathrm{ord}_P\,\Phi_{\phi^k,m}.$$

In characteristic p, Zieve [123] has given an explicit upper bound for the exponent e appearing in (c). See also [110, Theorem 2.28], and [49] for Hutz's generalization to higher dimensional maps. □

Remark 4.5. Here is how to formulate the theory of dynatomic polynomials in dimension greater than one. Let

$$\begin{aligned} \phi\colon X \to X &\quad \text{a self-morphism of a smooth projective variety,} \\ \Delta\colon X \to X \times X &\quad \text{the diagonal map,} \\ \Gamma_{\phi^n} \subset X \times X &\quad \text{the graph of } \phi^n. \end{aligned}$$

Then the analogue of $YF_n - XG_n$ is the 0-cycle $\Delta^*(\Gamma_{\phi^n})$, and the analogue of $\Phi_{\phi,n}$ is the 0-cycle

$$\Phi_{\phi,n} = \sum_{k|n} \mu(n/k)\Delta^*(\Gamma_{\phi^k}) \in Z_0(V).$$

Theorem 4.6 (Hutz [49–51]). $\Phi_{\phi,n}$ *is an effective* 0*-cycle, i.e., the points in the support of* $\Phi_{\phi,n}$ *appear with nonnegative multiplicity. Further, the points appearing in* $\Phi_{\phi,n}$ *with nonnegative multiplicity satisfy relations similar to those described in Theorem* 4.4.

4.2. Dynamical modular curves for $z^2 + c$

In this section we study the simplest nontrivial examples of discrete dynamical systems:

$$\boxed{\begin{array}{c}\textit{Quadratic Polynomials in One Variable}\\ \phi(z) = Az^2 + Bz + C, \quad A \neq 0.\end{array}}$$

There is always an affine map

$$f(z) = \alpha z + \beta$$

that transforms ϕ into

$$\phi_c(z) = \phi^f(z) = z^2 + c.$$

(This is working over a field of characteristic different from 2, of course.) In fancier terms, there is an injective map

$$\begin{array}{ccccc} \mathbb{A}^1 & \longrightarrow & \mathsf{M}_2 & \xrightarrow[\sim]{(\sigma_1,\sigma_2)} & \mathbb{A}^2, \\ c & \longmapsto & \langle \phi_c \rangle & \longmapsto & (2, 4c), \end{array}$$

where we computed the values of σ_1 and σ_2 in Proposition 2.58. Thus in $\mathsf{M}_2^1 \cong \mathbb{A}^2$, the polynomials form the line $\sigma_1 = 2$.

We want to study periodic points of $\phi_c(z)$ as c varies, so we treat c as a variable and write

$$\Phi_n(c; z) = \prod_{k|n} \left(\phi_c^n(z) - z\right)^{\mu(n,k)} \in \mathbb{Z}[c, z].$$

The roots of this *dynatomic polynomial* are pairs (c_0, z_0) such that z_0 is a point of formal period n for the map $\phi_{c_0}(z) = z^2 + c_0$.

Definition 4.7. The *dynamical modular curve* $Y_1(n) \subset \mathbb{A}^2$ is the affine curve

$$Y_1(n) : \Phi_n(y; z) = 0.$$

We also define the projective curve $X_1(n)$ to be the normalization of the projective closure of $Y_1(n)$, i.e., $X_1(n)$ is a smooth projective model for the affine curve $Y_1(n)$.

Remark 4.8. The curves $X_1(1)$, $X_1(2)$, and $X_1(3)$ have genus 0. This is easy for the first two, since

$$X_1(1) : z^2 - z + y = 0 \quad \text{and} \quad X_1(2) : z^2 + z + y + 1 = 0$$

are smooth conics in $\mathbb{P}^2$. Checking that $X_1(3)$ has genus 0 is messier, since

$$\deg_y \Phi_3(y; z) = 3 \quad \text{and} \quad \deg_z \Phi_3(y; z) = 6;$$

see [110, Example 4.9].

The curves $Y_1(n)$ and $X_1(n)$ are modular curves in the appropriate sense, i.e., they are (coarse) moduli spaces.

Proposition 4.9. *Let* $\operatorname{char}(K) \neq 2$.

(a) *The map*

$$\begin{array}{ccc} Y_1(n) & \longrightarrow & \dfrac{\left\{ (\phi, \alpha) : \begin{array}{c} \phi \in \overline{K}[z],\ \deg(\phi) = 2,\ \alpha \in \overline{K} \\ \alpha \textit{ has formal period } n \textit{ for } \phi \end{array} \right\}}{\mathrm{PGL}_2\textit{-isomorphism}} \\ (c, \alpha) & \longmapsto & (z^2 + c, \alpha), \end{array}$$

is a bijection of sets.

(b) *Let*

$$\lambda\colon T \longrightarrow \mathrm{Poly}_2 \times \mathbb{A}^1$$

be a family of degree 2 polynomials with a point of formal period n. In other words,

$$\lambda(t) = (A_t z^2 + B_t z + C_t, P_t)$$

such that P_t is a point of formal period n for $A_t z^2 + B_t z + C_t$. Then the map of sets

$$T \longrightarrow Y_1(n), \quad t \longmapsto \lambda(t),$$

is a morphism.

PROOF. See [110, Theorem 4.11]. □

Remark 4.10. Proposition (4.9) says that $Y_1(n)$ is a coarse moduli space. In fact, $Y_1(n)$ is a fine moduli space for $n \geq 2$. However, for $n = 1$ we observe that the map $f(z) = 1/z$ fixes the points $(0, \pm 1) \in Y_1(1)$ corresponding to the polynomial $\phi(z) = z^2$ and the fixed points $\pm 1 \in \mathrm{Fix}(\phi)$. The fact that $(0, \pm 1)$ are fixed by a nontrivial automorphism means that $Y_1(1)$ cannot be a fine moduli space.

On the other hand, an elementary case-by-case analysis shows that $\mathrm{Aut}(\phi_c) = \{z\}$ except for the map $\phi_0(z) = z^2$, whose automorphism group is $\{z, z^{-1}\}$. Further, $\pm 1 \in \mathrm{Fix}(\phi_0)$ are the only points fixed by $f(z) = z^{-1}$, and 0 cannot have formal period greater than 1 for ϕ_0 since its multiplier $\phi_0'(0) = 2$ is not a root of unity. It follows that $X_1(n)$ is a fine moduli space. (See also Remark 4.19 for similar dynamical modular curves $C_1(n)$ that are not fine moduli spaces, no matter how large one takes n.)

There is a natural action of $\mathbb{Z}/n\mathbb{Z}$ on $Y_1(n)$ via

$$\phi\colon Y_1(n) \longrightarrow Y_1(n), \quad (y; z) \longmapsto (y; z^2 + y) = \bigl(y, \phi_y(z)\bigr).$$

This action lifts to $X_1(n)$.

Definition 4.11. The curves $Y_0(n)$ and $X_0(n)$ are the quotients of $Y_1(n)$ and $X_1(n)$ via the action of ϕ. As noted earlier, the quotient of a variety by a finite set of automorphisms always exists.

The points of $Y_0(n)$ classify isomorphism classes of pairs

$$(\phi, \mathcal{O}), \quad \text{where} \begin{cases} \phi \text{ is a degree 2 polynomial,} \\ \mathcal{O} \text{ is an orbit of a point of formal period } n. \end{cases}$$

The isomorphism class of $(\phi, \mathcal{O})$ consists of all points $\bigl(\phi^f, f^{-1}(\mathcal{O})\bigr)$ taken over all invertible affine maps $f = \alpha z + \beta$.

There are projections which, from the moduli viewpoint, are defined by

$$Y_1(n) \longrightarrow Y_0(n), \quad (\phi, \alpha) \longmapsto \bigl(\phi, \mathcal{O}_\phi(\alpha)\bigr).$$

4.3. Irreducibility and genus formulas

The next result describes the geometry of the dynamical modular curves $Y_1(n)$, $X_1(n)$, and $X_0(n)$. It includes results of a number of authors. In particular, the smoothness of $Y_1(n)$ is due to Douady–Hubbard, and the irreducibility of $X_1(n)$ was proven using a variety of techniques by Bousch, Lau–Schleicher, and Morton. There is also a recent alternative proof of smoothness and irreducibility by Buff and Lei [24].

Theorem 4.12 (Bousch [21], Douady–Hubbard [35], Morton [83], Lau–Schleicher [63]).

(a) *The affine curve*

$$Y_1(n) : \Phi_n(y, z) = 0$$

is nonsingular.

(b) *The curves $X_0(n)$ and $X_1(n)$ are irreducible.*

(c) *The projection map*

$$X_1(n) \longrightarrow \mathbb{P}^1, \quad (\phi_c(z), \alpha) \longmapsto c,$$

is Galois with maximal Galois group (which is a wreath product).

(d) *Define*

$$\kappa(n) = \frac{1}{2} \sum_{k|n} \mu\left(\frac{n}{k}\right) 2^k.$$

Then

$$\operatorname{genus} X_1(n) = 1 + \frac{n-3}{2}\kappa(n) - \frac{1}{2} \sum_{\substack{m|n \\ m<n}} m\kappa(m)\varphi\left(\frac{n}{m}\right).$$

(There is a similar, but messier, formula for the genus of $X_0(n)$.)

PROOF SKETCH. We sketch Bousch's proof of (a) and (b), following notes of Michelle Manes. We note that since there is a covering map $X_1(n) \to X_0(n)$, it suffices in (b) to prove that $X_1(n)$ is irreducible.

(a) Let (z_0, c_0) be a point on the curve $Y_1(n)$. If

$$\left.\frac{\partial}{\partial z}\Phi_n(z, c_0)\right|_{z=z_0} \neq 0,$$

then (z_0, c_0) is a nonsingular point of $Y_1(n)$. So we are reduced to the case that

$$\left.\frac{\partial}{\partial z}\Phi_n(z, c_0)\right|_{z=z_0} = 0.$$

This means that z_0 appears as (at least) a double root of $\Phi_n(z, c_0)$. It follows that the corresponding multiplier $\lambda_{\phi_{c_0}}(z_0)$ equals 1 (cf. Proposition 2.31), so c_0 is a root of a hyperbolic component of the Mandelbrot set. But each hyperbolic component of the Mandelbrot set has a unique root (including multiplicity), since each such component is conformally isomorphic to the closed unit disk. Hence

$$\left.\frac{\partial}{\partial c}\Phi_n(z_0, c)\right|_{c=c_0} \neq 0,$$

so (z_0, c_0) is a nonsingular point of $Y_1(n)$.

(b) We will use two facts from classical complex dynamics.

(1) Let $c \in \mathbb{C}$ be a point that is not in the Mandelbrot set, i.e., such that

$$\lim_{n\to\infty} \phi^n(0) = \infty.$$

Then the Julia set $\mathcal{J}(\phi_c)$ of ϕ_c is contained in two disjoint open sets, say $\mathcal{J}(\phi_c) \subset \mathcal{U}_0 \cup \mathcal{U}_1$, and the dynamics of ϕ_c on its Julia set is conjugate to symbolic dynamics on two symbols via the itinerary map

$$I \colon \mathcal{J}(\phi_c) \longrightarrow \{0,1\}^{\mathbb{N}}, \quad z \longmapsto \bigl(I_n(z)\bigr)_{n\geq 0}.$$

Here $I_n(z)$ is determined by the condition $\phi_c^n(z) \in \mathcal{U}_{I_n(z)}$

(2) Again let c be outside the Mandelbrot set, let $z, z' \in \mathcal{J}(\phi_c)$ be distinct points of exact period n, and let $k = k(z, z')$ be the first index for which their itineraries differ. Then

$$z - z' \sim \pm(2\sqrt{-c})^{1-k}. \tag{4.2}$$

In particular, we have

$$\lim_{\substack{c \notin \mathcal{M} \\ z,z' \in \mathcal{J}(\phi_c) \cap \operatorname{Per}_n^{**}(\phi_c) \\ z \neq z'}} |z - z'| \cdot (2|c|^{1/2})^{k(z,z')-1} = 1.$$

We will also need a special case of Zsigmondy's theorem (this version is actually due to Bang).

Lemma 4.13 (Bang–Zsigmondy Theorem). *For all $n \geq 2$ with $n \neq 6$ there is a prime p such that*

$$p \mid 2^n - 1 \qquad \text{and} \qquad p \nmid 2^d - 1 \quad \text{for all } 1 \leq d < n.$$

(For $n = 6$, we have $3^2 \mid 2^6 - 1$ and $3^2 \nmid 2^d - 1$ for $d < 6$.)

PROOF. Exercise for the reader; or see [125]. □

We next show that it suffices to prove irreducibility over $\mathbb{Z}$.

Lemma 4.14. *If $\Phi_n(z, c)$ is irreducible in $\mathbb{Z}[z, c]$, then it is irreducible in $\mathbb{C}[z, c]$.*

PROOF. For any given $c_0 \in \mathbb{C}$, we can factor $\Phi_n(z, c_0)$ over $\mathbb{C}$. However, for c outside the Mandelbrot set, we can do this factorization in the ring of Laurent series using the variable $\gamma = 1/\sqrt{-c}$. More precisely, we can (formally) factor

$$\Phi_n(z, c) = (z - z_1)(z - z_2) \cdots (z - z_m)$$

with

$$z_i = \frac{a_{i,-1}}{\gamma} + a_{i,0} + a_{i,1}\gamma + a_{i,2}\gamma^2 + \cdots \in \gamma^{-1}\mathbb{Q}[\![\gamma]\!].$$

Suppose now that Φ_n factors, say

$$\Phi_n(z, c) = A(z, c)B(z, c) \quad \text{with } A, B \in \mathbb{C}[z, c].$$

Since $\Phi_n(z, c)$ is monic in z, we may assume the same for $A(z, c)$ and $B(z, c)$. Reordering the roots of Φ_n, we can write

$$A(z, c) = (z - z_1)(z - z_2) \cdots (z - z_r) \in \gamma^{-1}\mathbb{Q}[\![\gamma]\!][z],$$

and similarly for $B(z, c)$. Thus the coefficients of A and B, as polynomials in z, are in $\mathbb{C}[c] \cap \gamma^{-1}\mathbb{Q}[\![\gamma]\!]$, i.e., the coefficients are simultaneously:

- polynomials in c with complex coefficients,
- Laurent series in $(-c)^{-1/2}$ with rational coefficients.

Hence A and B are in $\mathbb{Q}[z, c]$. Thus $\Phi_n(z, c)$ factors in $\mathbb{Q}[z, c]$, and since the dynatomic polynomial $\Phi_n(z, c)$ has integer coefficients, Gauss's lemma tells us that it factors in $\mathbb{Z}[z, c]$. □

We now resume the proof of Theorem 4.12(b). From Lemma 4.14, it suffices to prove a contradiction from the assumption that $\Phi_n(z, c)$ factors as

$$\Phi_n(z, c) = A(z, c)B(z, c) \quad \text{with } A, B \in \mathbb{Z}[z, c] \text{ monic in } z.$$

From (a) we know that $Y_1(n)$ is nonsingular, so the affine curves $A(z,c) = 0$ and $B(z,c) = 0$ have no intersection points in $\mathbb{A}^2$, since any such intersection point would necessarily be a singular point of $Y_1(n)$. It follows that the polynomial

$$\operatorname{Resultant}_z\bigl(A(z,c), B(z,c)\bigr) \in \mathbb{C}[c]$$

has no complex roots, so it must be constant. On the other hand, using the relation (4.2) from earlier, we see that

$$\operatorname{Resultant}_z\bigl(A(z,c), B(z,c)\bigr) \sim \pm(2\sqrt{-c}\,)^t$$

for some integer t, so the fact that the resultant is constant implies that $t = 0$, and hence that $\operatorname{Resultant}_z(A, B) = \pm 1$.

We write

$$C_D(z) = \prod_{k|D} (z^k - 1)^{\mu(D/k)}$$

for the Dth cyclotomic polynomial. Then specializing the dynatomic polynomial $\Phi_n(z,c)$ to $c = 0$, we obtain the factorization

$$\Phi_n(z,0) = \prod_{d|n} \left(z^{2^d} - z\right)^{\mu(n/d)} = \prod_{D\in\mathcal{D}} C_D(z),$$

where

$$\mathcal{D} = \bigl\{k : k \mid 2^n - 1\bigr\} \smallsetminus \bigcup_{d|n,\ d\ne n} \bigl\{k : k \mid 2^d - 1\bigr\}.$$

Using the putative nontrivial factorization $\Phi_n(z,0) = A(z,0)B(z,0)$, we can write $\mathcal{D}$ as a disjoint union $\mathcal{D} = \mathcal{A} \cup \mathcal{B}$ so that there are factorizations

$$A(z,0) = \prod_{D\in\mathcal{A}} C_D(z) \quad\text{and}\quad B(z,0) = \prod_{D\in\mathcal{B}} C_D(z).$$

Further, since the cyclotomic polynomials are irreducible in $\mathbb{Q}[z]$, these give the complete factorizations of $A(z,0)$ and $B(z,0)$ into irreducible factors.

This gives a factorization of the resultant

$$1 = \bigl|\operatorname{Resultant}_z\bigl(A(z,0), B(z,0)\bigr)\bigr| = \prod_{(k,\ell)\in\mathcal{A}\times\mathcal{B}} \bigl|\operatorname{Resultant}_z\bigl(C_k(z), C_\ell(z)\bigr)\bigr|.$$

The cyclotomic resultants are integers, so we find that

$$\bigl|\operatorname{Resultant}_z\bigl(C_k(z), C_\ell(z)\bigr)\bigr| = 1 \quad \text{for all } (k,\ell) \in \mathcal{A}\times\mathcal{B}. \tag{4.3}$$

We now put various graph structures on $\mathcal{D}$ by specifying the pairs of points that are connected by an edge. For example, the graph structure $\mathcal{D}\times\mathcal{D}$ is the complete graph on $\mathcal{D}$ in which every vertex (point of $\mathcal{D}$) is connected to every other vertex.

Consider first the graph on $\mathcal{D}$ whose edges are given by the pairs

$$\mathcal{G}_1 = \bigl\{(k,\ell) \in \mathcal{G} : \bigl|\operatorname{Resultant}_z\bigl(C_k(z), C_\ell(z)\bigr)\bigr| > 1\bigr\}.$$

(By abuse of terminology, we will call $\mathcal{G}_1$ a graph.) It follows from (4.3) that in $\mathcal{G}_1$, there are no edges connecting a point in $\mathcal{A}$ to a point in $\mathcal{B}$, so the graph $\mathcal{G}_1$ is disconnected.

But then the graph

$$\mathcal{G}_2 = \bigl\{(k,\ell) \in \mathcal{G} : k \ne \ell,\ k \mid \ell,\ \bigl|\operatorname{Resultant}_z\bigl(C_k(z), C_\ell(z)\bigr)\bigr| > 1\bigr\}$$

is also disconnected, since the edges in $\mathcal{G}_2$ are a subset of the edges in $\mathcal{G}_1$. We now observe that if $k \ne \ell$ and $k \mid \ell$, then

$$\left|\operatorname{Resultant}_z\bigl(C_k(z), C_\ell(z)\bigr)\right| > 1 \iff \ell/k \text{ is a prime power.}$$

Hence the graph

$$\mathcal{G}_3 = \bigl\{(k,\ell) \in \mathcal{G} : k \ne \ell,\ k \mid \ell,\ \ell/k \text{ is a prime power}\bigr\}$$

is also disconnected, since $\mathcal{G}_3 \subset \mathcal{G}_2$.

The final part of the proof is to obtain a contradiction by showing that $\mathcal{G}_3$ is connected. Let $N = 2^n - 1$, and factor N as

$$N = p_1^{e_1} p_2^{e_2} \cdots p_r^{e_r}.$$

After possibly relabeling the primes, Lemma 4.13 tells us that

$$p_1^{e_1} \nmid 2^d - 1 \quad \text{for all } d \mid n \text{ with } d < n.$$

Hence if E is any divisor of N with $\operatorname{ord}_{p_1}(E) = e_1$, then the definition of $\mathcal{D}$ implies that $E \in \mathcal{D}$.

We now show that an arbitrary element $D \in \mathcal{D}$ is connected to the point $N \in \mathcal{D}$ in the graph $\mathcal{G}_3$, which will show that $\mathcal{G}_3$ is connected. We factor D as

$$D = p_1^{f_1} p_2^{f_2} \cdots p_r^{f_r} \quad \text{with } 0 \le f_i \le e_i.$$

For each $0 \le j \le r$, let

$$D_j = p_1^{e_1} p_2^{e_2} \cdots p_j^{e_j} \cdot p_{j+1}^{f_{j+1}} \cdots p_r^{f_r},$$

where by convention we set $D_0 = D$, and we note that $D_r = N$. For $j \ge 1$ we have $\operatorname{ord}_{p_1}(D_j) = e_j$, so our earlier discussion implies that $D_j \in \mathcal{D}$, and of course $D_0 = D \in \mathcal{D}$ by assumption. Further, we have

$$\frac{D_{j+1}}{D_j} = p_j^{e_j - f_j},$$

so either $D_{j+1} = D_j$, or D_{j+1}/D_j is a prime power. Letting $j_1 < j_2 < \cdots < j_t$ be the values of j satisfying $D_{j+1} \ne D_j$, we see that

$$\bigl(D_{j_u}, D_{j_{u+1}}\bigr) \in \mathcal{G}_3 \quad \text{for all } 1 \le u < t,$$

and that the list of pairs

$$\bigl(D_{j_1}, D_{j_2}\bigr),\ \bigl(D_{j_2}, D_{j_3}\bigr), \ldots,\ \bigl(D_{j_{t-1}}, D_{j_t}\bigr) \in \mathcal{G}_3$$

gives a path connecting D to N. This proves that $\mathcal{G}_3$ is connected, contradicting our earlier proof that $\mathcal{G}_3$ is disconnected. Hence $\Phi_n(z,c)$ is irreducible, which completes the proof of Theorem 4.12. □

The genera of $X_0(n)$ and $X_1(n)$ grow quite rapidly, as indicated in Table 4.1.

TABLE 4.1

n	1	2	3	4	5	6	7
genus $X_1(n)$	0	0	0	2	14	34	124
genus $X_0(n)$	0	0	0	0	2	4	16

4.4. Rational points on dynamical modular curves

Recall that some results are known concerning rational periodic points of small period for $\phi_c(z) = z^2 + c$; see Table 1.1 on page 9. For example, the fact that

$$X_1(n) \cong \mathbb{P}^1 \quad \text{for } n = 1, 2, 3$$

implies that for these n, there are many values of $c \in \mathbb{Q}$ for which $\phi_c(z)$ has a rational period point of exact period n.

Morton [82] observed that $X_1(4)$ is isomorphic to an elliptic modular curve that has finitely many rational points, and those rational points correspond to points in $X_1(4) \smallsetminus Y_1(4)$. Hence for $c \in \mathbb{Q}$, the map $\phi_c(z)$ never has rational points of exact period 4. For $X_1(5)$, Flynn, Poonen, and Schaefer [42] found all points on $X_0(5)$, and Stoll [115] did the same for $X_0(6)$ under the assumption that the Birch–Swinnerton-Dyer conjecture holds for the four-dimensional Jacobian variety $J_0(6)$ of $X_0(6)$. Using these results, they prove that $Y_1(n)(\mathbb{Q}) = \varnothing$ for $n = 5$ and 6. But for $n \geq 7$, it seems difficult to do arithmetic calculations on $X_0(n)$ and/or $J_0(n)$, since even $J_0(7)$ has dimension 16.

4.5. Other dynamical modular curves

There are many other interesting 1-dimensional families of rational maps. For example, much of the preceding theory goes over to the family

$$\phi(z) = z^d + c.$$

In order to describe another family, we recall that the automorphism group of a rational map is the group

$$\operatorname{Aut}(\phi) = \{f \in \mathrm{PGL}_{n+1} : \phi^f = \phi\}.$$

Since

$$\operatorname{Aut}(\phi^f) = \operatorname{Aut}(\phi)^f,$$

the automorphism group is well-defined (as an abstract group) for $\phi \in \mathsf{M}_d^n$. For example, one can show that for maps $\phi \in \mathsf{M}_2^1 \cong \mathbb{A}^2$ of degree two on $\mathbb{P}^1$, we have

$$\operatorname{Aut}(\phi) = 1, \quad \mathbb{Z}/2\mathbb{Z}, \quad \text{or} \quad \mathcal{S}_3.$$

Proposition 4.15. *The set*

$$C = \bigl\{\phi \in \mathsf{M}_2 : \operatorname{Aut}(\phi) \neq 1\bigr\} \subset \mathbb{A}^2$$

is a cuspidal cubic curve in $\mathbb{A}^2$ with the cusp corresponding to the unique map in M_2 having $\operatorname{Aut}(\phi) = \mathcal{S}_3$.

PROOF SKETCH. Looking at the nontrivial finite subgroups of PGL_2, we see that they all contain an element of order 2, so for any $\psi \in C$ we find that $\operatorname{Aut}(\phi)$ contains a nontrivial automorphism of order two. This automorphism has two fixed points. Moving the fixed points to 0 and ∞ and scaling, we find that the points of C correspond to maps of the form

$$\psi_b(z) = \frac{z}{z^2 + b}, \qquad b \neq 0,$$

with automorphism group

$$\operatorname{Aut}(\psi_b) \supset \{\pm z\}.$$

The fixed points of ψ_b are

$$\operatorname{Fix}(\psi_b) = \bigl\{0, \pm\sqrt{1 - b}\,\bigr\},$$

and the multipliers of the fixed points are

$$\psi'(0) = \frac{1}{b} \quad \text{and} \quad \psi'\big(\pm\sqrt{1-b}\big) = 2b - 1. \tag{4.4}$$

Hence

$$\sigma_1(\psi_b) = 4b - 2 + \frac{1}{b} \quad \text{and} \quad \sigma_2(\psi_b) = 4b^2 - 4b + 5 - \frac{2}{b}.$$

Eliminating b from these two equations, e.g., by computing the resultant

$$\operatorname{Resultant}_b(b\sigma_1 - 4b^2 + 2b - 1, b\sigma_2 - 4b^3 + 4b^2 - 5b + 2),$$

gives the following relation that σ_1 and σ_2 must satisfy:

$$2\sigma_1^3 + \sigma_1^2\sigma_2 - \sigma_1^2 - 8\sigma_1\sigma_2 + 12\sigma_1 - 4\sigma_2^2 + 12\sigma_2 - 36 = 0. \tag{4.5}$$

This defines the curve in $\mathsf{M}_2 \cong \mathbb{A}^2$ associated to maps with a nontrivial automorphism group. Taking partial derivatives, a calculation shows that this cubic curve has exactly one singular point, namely a node at the point $(\sigma_1, \sigma_2) = (-6, 12)$.

Finally one checks that

$$\operatorname{Aut}(\phi) = \mathcal{S}_3 \iff \phi(z) \text{ is conjugate to } z^{-2}.$$

(Note that $\operatorname{Aut}(z^{-2}) \cong \mathcal{S}_3$ is generated by the maps z^{-1} and ρz, where ρ is a primitive cube root of unity.) The fixed points of the map z^{-2} are the three cube roots of unity, each of which has multiplier -2. Thus $\sigma_1(z^{-2}) = -6$ and $\sigma_2(z^{-2}) = 12$, so z^{-2} corresponds to the singular point on the curve (4.5). □

Remark 4.16. Polynomials $z^2 + c$ are nice because they have only one critical point whose orbit is interesting. The maps $\psi_b(z)$ have two critical points, but the critical points are interchanged by $\operatorname{Aut}(\psi_b)$, so they can be analyzed by following the orbit of either critical point.

Now we add level structure.

Definition 4.17.

$$C_1(n) = \left\{(b, \beta) : \beta \text{ is a point of formal period } n \text{ for } \psi_b(z) = \frac{z}{z^2 + b}\right\}.$$

The structure of these curves has been studied by Manes. Here's a special case of one of her results.

Theorem 4.18 (Manes [68]). *$C_1(2m)$ is reducible for all $m \geq 2$.*

PROOF SKETCH. For a generic b, consider the collection of orbits

$$\{\mathcal{O}_{\psi_b}(\beta) : \beta \in \operatorname{Per}^*_{2m}(\psi_b)\},$$

where the notation means β is a point of formal period $2m$. Let $g(z) = -z$. Note that $\operatorname{Aut}(\psi_b) = \{1, g\}$ acts on the set of points of formal period $2m$. Since g fixes ψ_b, we have

$$g\big(\mathcal{O}_{\psi_b}(\beta)\big) = \mathcal{O}_{\psi_b^{g^{-1}}}\big(g(\beta)\big) = \mathcal{O}_{\psi_b}\big(g(\beta)\big).$$

There are two possibilities:

$$\mathcal{O}_{\psi_b}\big(g(\beta)\big) = \mathcal{O}_{\psi_b}(\beta) \quad \text{for one (all) } \beta \in \operatorname{Per}^*_n(\psi_b),$$
$$\mathcal{O}_{\psi_b}\big(g(\beta)\big) \neq \mathcal{O}_{\psi_b}(\beta) \quad \text{for one (all) } \beta \in \operatorname{Per}^*_n(\psi_b).$$

One then checks that these two cases break the modular curve into two components. (Note that in the first case, the group $\operatorname{Aut}(\psi_b) = \mathbb{Z}/2\mathbb{Z}$ acts nontrivially on $\mathcal{O}_{\psi_b}(\beta)$, which is only possible if $\#\mathcal{O}_{\psi_b}(\beta)$ is even. That's why the theorem applies to points of even period.) $\square$

Remark 4.19. Although we have added level structure to form the curve $C_1(n)$, it turns out that this is not enough to make $C_1(n)$ into a fine moduli space. The problem is that there are always pairs $(b,\beta) \in C_1(n)$ that admit an automorphism. More precisely, we noted earlier (4.4) that the multipliers of the fixed points of ψ_b are

$$\psi_b'(0) = \frac{1}{b} \quad \text{and} \quad \psi_b'\bigl(\pm\sqrt{1-b}\,\bigr) = 2b - 1.$$

If we choose b to be a primitive nth-root of unity, then $\psi_b'(0) = b^{-1}$ is also a primitive nth-root of unity, so the fixed point 0 is a point of formal period n. Thus

$$(\psi_b, 0) \in C_1(n).$$

But the pair $(\psi_b, 0)$ is fixed by the automorphism $z \mapsto -z$, so $C_1(n)$ has a point fixed by a nontrivial automorphism, and hence $C_1(n)$ is not a fine moduli space. (This remark is due to Michelle Manes and Alon Levy.)

CHAPTER 5

Canonical Heights

This chapter is devoted to theory of height functions. The following bullet points give a flavor of what height functions are and why they're useful in the study of arithmetic dynamics.

• Height functions give a way of measuring the arithmetic complexity of objects, such as points, maps, varieties,... .

• Height functions provide a tool for translating geometric facts into arithmetic facts.

• Height functions transform functorially with respect to maps. This leads to the definition of canonical heights, whose transformation properties are particularly nice.

• Canonical height functions can be decomposed as sums of local height functions, one local height for each absolute value. These canonical local heights, which are also known as Green functions, are analytic (in an appropriate sense) and can be used to study complex and p-adic dynamics.

5.1. Heights and projective space

The theory of height functions is covered in many books, including for example [17, 48, 62, 110, 111].

The basic set-up is:

K a global field (a number field or a one-dimensional function field),

M_K a complete set of inequivalent normalized absolute values on K.

Definition 5.1. The (*absolute logarithmic*) *Weil height* on $\mathbb{P}^N$ is

$$h\colon \mathbb{P}^N(\overline{K}) \longrightarrow \mathbb{R},$$

$$h([x_0,\dots,x_N]) = \sum_{v\in M_L} \frac{[L_v:\mathbb{Q}_v]}{[L:\mathbb{Q}]} \log\max\{|x_0|_v,\dots,|x_N|_v\},$$

where L/K is any extension such that $[x_0,\dots,x_N]\in\mathbb{P}^N(L)$. For numbers $x\in\overline{K}$, we write $h(x)$ for $h([1,x])$.

The following finiteness result is of fundamental importance in virtually all applications of height functions to number theory, arithmetic geometry, and arithmetic dynamics.

Proposition 5.2. *Fix real numbers A and B. If K is a number field, then*

$$\{P\in\mathbb{P}^N(\overline{K}) : h(P)\le A \text{ and } [K(P):K]\le B\} \tag{5.1}$$

is a finite set.

If K is a function field, then the set (5.1) can be identified with the points of a "finite dimensional variety defined over the field of constants of K." (This is not meant to be precise.)

Example 5.3. Let $K = k(T)$ be the field of rational functions in one variable. Write a point $P \in \mathbb{P}^N(K)$ as

$$P = [x_0(T), \ldots, x_N(T)] \quad \text{with } x_0, \ldots, x_N \in k[T] \text{ and } \gcd(x_0, \ldots, x_N) = 1.$$

Then

$$h(P) = \max_{0 \le i \le N} \deg x_i(T).$$

Bounding $h(P)$ gives a union of finite dimensional affine spaces over k, since bounding $\deg x_i(T)$ gives a finite number of freely varying coefficients.

Another important property of height functions is that they translate geometric information into arithmetic information, as in the following result.

Proposition 5.4. *Let $\phi \in \operatorname{Hom}_d^N(\overline{K})$. Then*

$$h(\phi(P)) = dh(P) + O(1) \quad \text{for all } P \in \mathbb{P}^N(\overline{K}), \tag{5.2}$$

where the $O(1)$ depends on ϕ, but not on P.

More precisely, use the projective embedding $\operatorname{Hom}_d^N \subset \mathbb{P}^M$ as described in Section 1.4 and define $h(\phi)$ to be the height of the point $\phi \in \mathbb{P}^M$. Then

$$h(\phi(P)) = dh(P) + O_{N,d}(h(\phi) + 1) \quad \text{for all } P \in \mathbb{P}^N(\overline{K}), \tag{5.3}$$

where now the big-O constant depends only on N and d.

Proof. The upper bound $h(\phi(P)) \le dh(P) + O(1)$ is an easy consequence of the triangle inequality. The lower bound uses the Nullstellensatz, as well as the triangle inequality. For details, see for example [111, Theorem VIII.5.6] or [110, Theorem 3.11]. The proof of the stronger estimate (5.3) follows along the same lines, but requires a version of the Nullstellensatz that shows that the coefficients of the polynomials used to form linear combinations are polynomials in the coefficients of the original polynomials. □

5.2. Dynamical canonical heights

The height estimates (5.2) and (5.3) are quite powerful, but the big-O terms that appear can be difficult to control, especially as ϕ is iterated. The next result shows how a limiting process can be used to eliminate these terms.

Theorem 5.5 (Tate). *Let $\phi \in \operatorname{Hom}_d^N(\overline{K})$. Then the limit*

$$\hat{h}_\phi(P) = \lim_{n \to \infty} \frac{1}{d^n} h(\phi^n(P)) \tag{5.4}$$

exists and satisfies:

(a) *$\hat{h}_\phi(P) = h(P) + O(1)$ for all $P \in \mathbb{P}^N(\overline{K})$. (The $O(1)$ depends on ϕ, but as in Proposition 5.4, it may be replaced by $O_{N,d}(h(\phi) + 1)$.)*

(b) *$\hat{h}_\phi(\phi(P)) = d\hat{h}_\phi(P)$ for all $P \in \mathbb{P}^N(\overline{K})$.*

The function $\hat{h}_\phi : \mathbb{P}^N(\overline{K}) \to \mathbb{R}$ is uniquely determined by properties (a) and (b).

PROOF. We prove that the limit (5.4) exists by proving that the sequence is Cauchy. Let $n > m \geq 0$ be integers. Proposition 5.4 tells us that there is a constant C such that

$$\bigl|h(\phi(Q)) - dh(Q)\bigr| \leq C \qquad \text{for all } Q \in \mathbb{P}^N(\overline{K}). \tag{5.5}$$

We apply inequality (5.5) with $Q = \phi^{i-1}(P)$ to the telescoping sum

$$\begin{aligned} \left|\frac{1}{d^n}h(\phi^n(P)) - \frac{1}{d^m}h(\phi^m(P))\right| &= \left|\sum_{i=m+1}^{n} \frac{1}{d^i}\Bigl(h(\phi^i(P)) - dh(\phi^{i-1}(P))\Bigr)\right| \\ &\leq \sum_{i=m+1}^{n} \frac{1}{d^i}\bigl|h(\phi^i(P)) - dh(\phi^{i-1}(P))\bigr| \\ &\leq \sum_{i=m+1}^{n} \frac{C}{d^i} \\ &\leq \frac{C}{d^m(d-1)}. \end{aligned} \tag{5.6}$$

The inequality (5.6) clearly implies that

$$\left|\frac{1}{d^n}h(\phi^n(P)) - \frac{1}{d^m}h(\phi^m(P))\right| \to 0 \quad \text{as } m, n \to \infty,$$

so the sequence $d^{-n}h(\phi^n(P))$ is Cauchy and the limit (5.4) exists.

In order to prove (a), we take $m = 0$ in (5.6), which yields

$$\left|\frac{1}{d^n}h(\phi^n(P)) - h(P)\right| \leq \frac{C}{d-1}.$$

Next we let $n \to \infty$ to obtain

$$\bigl|\hat{h}_\phi(P) - h(P)\bigr| \leq \frac{C}{d-1},$$

which is (a) with an explicit value for the $O(1)$ constant.

The proof of (b) is a simple computation using the definition of $\hat{h}_\phi$,

$$\hat{h}_\phi(\phi(P)) = \lim_{n\to\infty} \frac{1}{d^n}h\bigl(\phi^n(\phi(P))\bigr) = \lim_{n\to\infty} \frac{d}{d^{n+1}}h\bigl(\phi^{n+1}(P))\bigr) = d\hat{h}_\phi(P).$$

Finally, to prove uniqueness, suppose that $\hat{h}'_\phi : \mathbb{P}^N(\overline{K}) \to \mathbb{R}$ also has properties (a) and (b). Then the difference $g = \hat{h}_\phi - \hat{h}'_\phi$ satisfies

$$g(P) = O(1) \quad \text{and} \quad g(\phi(P)) = dg(P).$$

These formulas hold for all points $P \in \mathbb{P}^N(\overline{K})$, so

$$d^n g(P) = g(\phi^n(P)) = O(1) \quad \text{for all } n \geq 0.$$

In other words, the quantity $d^n g(P)$ is bounded as $n \to \infty$, which can happen only if $g(P) = 0$. This proves that $\hat{h}_\phi(P) = \hat{h}'_\phi(P)$, so $\hat{h}_\phi$ is unique. □

Corollary 5.6. *Let* $\phi \in \mathrm{Hom}_d^N(K)$.

(a) **(Northcott 1950 [91]).** $\mathrm{PrePer}(\phi)$ *is a set of bounded height.*

(b) *Let K is a number field and $P \in \mathbb{P}^N(\overline{K})$. Then*

$$\hat{h}_\phi(P) = 0 \iff P \in \operatorname{PrePer}(\phi).$$

(*See Section* 5.3 *for a discussion of $\hat{h}_\phi(P) = 0$ over function fields.*)

PROOF. We have

$$\begin{aligned} P \in \operatorname{PrePer}(\phi) &\implies h\big(\phi^n(P)\big) \text{ is bounded} \\ &\implies \hat{h}_\phi(P) = \lim_{n\to\infty} d^{-n} h\big(\phi^n(P)\big) = 0 \\ &\implies h(P) = \hat{h}_\phi(P) + O(1) = O(1). \end{aligned}$$

This proves (a) and one direction of (b).

For the other direction, we have

$$\begin{aligned} \hat{h}_\phi(P) = 0 \implies h\big(\phi^n(P)\big) &= \hat{h}_\phi\big(\phi^n(P)\big) + O(1) \\ &= d^n \hat{h}_\phi(P) + O(1) \\ &= O(1) \\ \implies \mathcal{O}_\phi(P) &\text{ is a set of bounded height.} \end{aligned}$$

Further, it is clear that $\mathcal{O}_\phi(P)$ is defined over a field of bounded degree, since for example

$$\mathcal{O}_\phi(P) \subset \mathbb{P}^N\big(K(P)\big).$$

Thus $\mathcal{O}_\phi(P)$ is a set of bounded degree and bounded height, so if K is a number field, then Proposition 5.2 tells us that $\mathcal{O}_\phi(P)$ is a finite set. Hence $P \in \operatorname{PrePer}(\phi)$. $\square$

The canonical height can be used in various ways to define interesting functions on Hom_d^N and M_d^N. We mention two pairings that have recently been studied. (See also Definition 6.24 for another use of canonical heights.)

Theorem 5.7 (Kawaguchi, Silverman [59]). *For $\phi, \psi \in \operatorname{Hom}_d^N(\overline{\mathbb{Q}})$, define*

$$[\phi,\psi]_{\mathrm{KS}} = \sup_{P\in\mathbb{P}^N(\overline{K})} \big|\hat{h}_\phi(P) - \hat{h}_\psi(P)\big|.$$

Then for fixed ψ, the map

$$\operatorname{Hom}_d^N \longrightarrow \mathbb{R}, \quad \phi \longmapsto [\phi,\psi]_{\mathrm{KS}},$$

is a height function on Hom_d^N. In other words, if we let let

$$h\colon \operatorname{Hom}_d^N(\overline{\mathbb{Q}}) \subset \mathbb{P}^M(\overline{\mathbb{Q}}) \longrightarrow \mathbb{R}$$

be the Weil height on $\mathbb{P}^M(\overline{\mathbb{Q}})$, then there are positive constants $c_1,\dots,c_4$, depending on N, d, and ψ, such that

$$c_1 h(\phi) - c_2 \le [\phi,\psi]_{\mathrm{KS}} \le c_3 h(\phi) + c_4 \quad \textit{for all } \phi \in \operatorname{Hom}_d^N(\overline{\mathbb{Q}}).$$

Theorem 5.8 (Petsche, Szpiro, Tucker [96]). *For $\phi,\psi \in \operatorname{Hom}_d(\overline{\mathbb{Q}})$, the limit*

$$[\phi,\psi]_{\mathrm{AZ}} = \lim_{\substack{P\in\mathbb{P}^1(\overline{K}) \\ \hat{h}_\phi(P)\to 0}} \hat{h}_\psi(P).$$

exists, is symmetric, i.e., $[\phi,\psi]_{\mathrm{AZ}} = [\psi,\phi]_{\mathrm{AZ}}$, and satisfies

$$[\phi,\psi]_{\mathrm{AZ}} = \lim_{n\to\infty} \frac{1}{(\deg\psi)^n + 1} \sum_{P\in\operatorname{Per}_n(\psi)} \hat{h}_\phi(P).$$

(*The authors call* $[\cdot,\cdot]_{\mathrm{AZ}}$ *the* Arakelov–Zhang pairing.)

The following properties are equivalent:

(a) $[\phi,\psi]_{\mathrm{AZ}}=0$.
(b) $\hat{h}_\phi=\hat{h}_\psi$.
(c) $\mathrm{PrePer}(\phi)=\mathrm{PrePer}(\psi)$.
(d) $\#\bigl(\mathrm{PrePer}(\phi)\cap\mathrm{PrePer}(\psi)\bigr)=\infty$.
(e) $\liminf_{P\in\mathbb{P}^1(\overline{\mathbb{Q}})}\bigl(\hat{h}_\phi(P)+\hat{h}_\psi(P)\bigr)=0$.

Remark 5.9. The equivalence of (a) and (e) in Theorem 5.8 also follows from Zhang's successive minima theorem [122, Theorem 1.10], while the equivalence of properties (c), (d), and (e) was proven by Mimar [81] using Arakelov theory. Chambert-Loir and Thuillier [27, §6] independently proved more general results on the equidistribution of small points on varieties with semipositive adelic metrics.

The KS and AZ pairings are not PGL_{n+1} invariant, so they give pairings only on the parameter space Hom_d^n, not on the moduli space M_d^n. This suggests defining invariant versions, for example

$$[\phi,\psi]'_{\mathrm{KS}}=\inf_{f,g\in\mathrm{PGL}_{n+1}(\overline{\mathbb{Q}})}[\phi^f,\psi^g]_{\mathrm{KS}},$$

and similarly for $[\phi,\psi]'_{\mathrm{AZ}}$. Then $[\cdot,\cdot]'_{\mathrm{KS}}$ and $[\cdot,\cdot]'_{\mathrm{AZ}}$ define pairings

$$\mathsf{M}_d^N\times\mathsf{M}_d^N\longrightarrow\mathbb{R}.$$

What can one say about them? For example, is the following true?

Question 5.10.

$$[\phi,\psi]'_{\mathrm{KS}}=0\quad\Longleftrightarrow\quad[\phi^f,\psi^g]_{\mathrm{KS}}=0\text{ for some }f,g\in\mathrm{PGL}_{n+1}(\overline{K})?$$

5.3. Canonical height zero over function fields

A key property of the Weil height on $\mathbb{P}^n(\overline{\mathbb{Q}})$ is that there are only finitely many points of bounded degree and bounded height. In particular, for a given number field K, there are only finitely many points in $\mathbb{P}^n(K)$ of bounded height. The analogous statement is not true for function fields whose constant field is infinite. For example, the height of a point

$$P=\bigl[x_0(T),\dots,x_n(T)\bigr]\in\mathbb{P}^n\bigl(k(T)\bigr)$$

whose homogeneous coordinates are polynomials with $\gcd(x_0,\dots,x_n)=1$ is simply

$$h(P)=\max_{0\le i\le n}\deg x_i(T).$$

If the constant field k is infinite, then there are clearly infinitely many points of bounded height. This in turn means that it is possible for there to be infinitely many points of canonical height 0, so it is no longer true in complete generality that $\hat{h}_\phi(P)=0$ implies that P is preperiodic for ϕ.

However, if one rules out the cases in which ϕ and P essentially come from maps defined over the constant field k, then appropriately formulated versions of this equivalence are true.

Theorem 5.11. *Let k be an algebraically closed field of characteristic* 0, *let K be the function field of a smooth projective curve defined over k, let $\phi\in\mathrm{Hom}_d^N(K)$, and let $P\in\mathbb{P}^N(\overline{K})$.*

(a) **(Baker [2], Benedetto [13]).** *If $N = 1$ and ϕ is not isotrivial, then*

$$\hat{h}_\phi(P) = 0 \quad \Longleftrightarrow \quad P \in \operatorname{PrePer}(\phi).$$

(b) **(Chatzidakis–Hrushovski [28]).** *Let N be arbitrary. Suppose that the dynamical system $(\phi, \mathbb{P}^N)$ does not admit a dominant rational map to a lower dimensional (but not zero-dimensional) dynamical system, and further assume that $(\phi, \mathbb{P}^N)$ is not rationally isotrivial, i.e., it is not birational to a dynamical system defined over k. Then there is a proper Zariski closed set $W \subset \mathbb{P}^N$ such that for all $P \in (\mathbb{P}^N \smallsetminus W)(\overline{K})$,*

$$\hat{h}_\phi(P) = 0 \iff P \in \operatorname{PrePer}(\phi).$$

PROOF. (a) This was proved for polynomial maps by Benedetto [13], and then for rational maps by Baker [2]. Both proofs use p-adic methods.

(b) The proof by Chatzidakis and Hrushovski [28] uses techniques from model theory and is much more general. Since model theoretic methods are not familiar to many people working in dynamics, we state their general result and discuss how it implies Theorem 5.11(b). This discussion follows notes of Tom Scanlon (private communication).

Before stating the result of Chatzidakis and Hrushovski, we need some definitions.

Definition 5.12. An algebraic dynamical system (V, ϕ) is *rationally imprimitive* if there is an algebraic dynamical system (Z, ψ) with $0 < \dim Z < \dim V$ and a dominant rational map

$$\lambda : V \dashrightarrow Z \quad \text{satisfying } \lambda \circ \phi = \psi \circ \lambda.$$

The system (V, ϕ) is *rationally primitive* if it is not rationally imprimitive.

Definition 5.13. Let K/k be a regular extension of fields, and let V/K be an algebraic variety. A subset $Y \subset V(K)$ is said to be *limited* if there is an algebraic variety W/k and a rational map $\mu : W \times_k K \to V$ such that

$$Y \subset \mu\bigl(W(k)\bigr).$$

Definition 5.14. Let K/k be a regular extension of fields, and let (V, ϕ) be an algebraic dynamical system defined over K. We say that (V, ϕ) *contructibly descends to k* if there is an algebraic dynamical system (W, ψ) defined over k and a rational map

$$\nu : V \dashrightarrow W \times_k K \quad \text{satisfying } \nu \circ \phi = \psi \circ \nu$$

such that ν induces a generically bijective map from $V(\overline{K})$ to $W(\overline{K})$. In other words, there are Zariski dense constructible subsets $V' \subset V$ and $W' \subset W$ such that $\nu : V'(\overline{K}) \to W'(\overline{K})$ is a bijection.

Theorem 5.15 (Chatzidakis–Hrushovski [28]). *Let K/k be a finitely generated regular extension of fields. Let (V, ϕ) be a rationally primitive algebraic dynamical system defined over K, and assume that (V, ϕ) does not construtibly descend to k. Let Y be a limited subset of $V(K)$. Then there is an $n = n(Y)$ and a proper Zariski closed subset $W \subsetneq V$ defined over K with the following property:*

$$\bigl\{P, \phi(P), \ldots, \phi^n(P)\bigr\} \not\subset Y \quad \text{for all } P \in V(K) \smallsetminus W(K).$$

In order to relate this to heights, let k be an algebraically closed field and K/k the the function field of a smooth curve defined over k. Further let V/K be a quasi-projective variety with a given projective embedding, so we can define the height of points in $V(K)$. Then for any B, the set

$$\{P \in V(K) : h(P) \le B\}$$

is a limited set. Further, if $\operatorname{char}(k) = 0$, then (V, ϕ) constructibly descends to k if and only if (V, ϕ) is birational to an algebraic dynamical system defined over k. (This is no longer true if $\operatorname{char}(k) = p > 0$.)

So in terms of points of canonical height zero, Theorem 5.15 says the following. Let K/k be as above, and let (V, ϕ) be an algebraic dynamical system defined over K that satisfies the following two conditions:

- (V, ϕ) is rationally primitive, i.e., it does not admit a dominat rational map to a lower (but not zero) dimensional dynamical system.
- (V, ϕ) does not constructably descend to k. In the case that $\operatorname{char}(k) = 0$, this means that (V, ϕ) is not rationally isomorphic to a dynamcal system defined over k.

Then the set

$$\{P \in V(K) : \hat{h}_\phi(P) = 0\}$$

is not Zariski dense in V. Thus counterexamples to the implication

$$\hat{h}_\phi(P) = 0 \implies P \text{ is preperiodic}$$

come from proper sub-dynamical systems that are themselves constructibly isotrivial.

The proof is as in the proof of Corollary 5.6. Suppose that $\hat{h}_\phi(P) = 0$. Since $\hat{h}_\phi - h = O(1)$ and $\hat{h}_\phi \circ \phi^n = d^n \hat{h}_\phi$, it follows that the orbit $\mathcal{O}_\phi(P)$ is a set of bounded height. Hence there is a B so that $\mathcal{O}_\phi(P)$ is contained in the set $\{Q \in V(K) : h(Q) \le B\}$. (Note that B depends only on ϕ; it is independent of the point P.) It follows from Theorem 5.15 that P is in a proper Zariski closed set W that depends only on the dynamical system (ϕ, V). □

Finally, we note that Bhatnagar and Szpiro have announced the following strong result for points of canonical height zero over function fields. Their proof is primarily algebro-geometric in nature and uses Hilbert schemes in a fundamental way.

Theorem 5.16 (Bhatnagar–Szpiro [16]; see also [15]). *Let $K = k(C)$ be the function field of a smooth projective connected curve C over an algebraically closed field k of characteristic zero. Let Y/K be a projective variety, let $\phi\colon Y \to Y$ be a self-morphism of Y defined over K, and let $D \in \operatorname{Div}(Y) \otimes \mathbb{Q}$ be an ample divisor with the property that $\phi^* D \sim rD$ for some $r > 1$. Then the set*

$$\{P \in Y(\overline{K}) : \hat{h}_{\phi,D}(P) = 0\}$$

is a finite union of the constant points of closed irreducible preperiodic isotrivial subvarieties of Y.

Remark 5.17. Ingram [55] has proven a result about points of canonical height zero for Hénon maps

$$\phi(x, y) = \big(y, x + f(y)\big) \colon \mathbb{A}^2 \longrightarrow \mathbb{A}^2$$

over function fields. (Note that ϕ although extends to a rational map $\mathbb{P}^2 \dashrightarrow \mathbb{P}^2$, it does not extend to a morphism.)

5.4. Local heights and Green functions

An important tool for studying canonical heights is a decomposition as a sum of local heights. This is most easily accomplished for polynomials, so we start with that case.

Definition 5.18. Let $\phi(z) \in K[z]$ be a polynomial of degree $d \geq 2$, and let $v \in M_{\overline{K}}$. The associated *v-adic local canonical height* (or *Green function*) is the map

$$\hat{g}_{\phi,v} : \mathbb{A}^1(\overline{K}_v) \longrightarrow \mathbb{R}$$

defined by the limit

$$\hat{g}_{\phi,v}(\alpha) = \lim_{n\to\infty} \frac{1}{d^n} \log^+ \bigl|\phi^n(\alpha)\bigr|_v.$$

Here $\log^+(t) = \max\bigl\{\log(t), 0\bigr\}$. We also formally set $\hat{g}_{\phi,v}(\infty) = \infty$.

We state a few of the basic properties of local height functions.

Proposition 5.19. *Let $\phi(z) \in K[z]$ be a polynomial of degree $d \geq 2$.*

(a) $\hat{g}_{\phi,v}\bigl(\phi(\alpha)\bigr) = d\hat{g}_{\phi,v}(\alpha)$.

(b) *The function*

$$\mathbb{A}^1(\overline{K}_v) \longrightarrow \mathbb{R}, \qquad z \longmapsto \hat{g}_{\phi,v}(z) - \log^+ |z|_v,$$

extends to a bounded function on $\mathbb{P}^1(\overline{K}_v)$.

(c) *Let $v \in M_{\overline{K}}$ be a non-Archimedean absolute value at which ϕ has good reduction, which in this case means the coefficients of ϕ are v-integral and the leading coefficient is a v-adic unit. Then*

$$\hat{g}_{\phi,v}(\alpha) = \log^+ |\alpha|_v \quad \textit{for all } \alpha \in \mathbb{A}^1(\overline{K}_v).$$

(d) *Let $v \in M_{\overline{K}}$ be an Archimedean absolute value and fix an isomorphism $\overline{K}_v \cong \mathbb{C}$. Then $g_{\phi,v}$ is a classical Green function relative to the filled Julia set $\mathcal{J}^{\mathrm{f}}(\phi)$ of ϕ. In other words,*

$$g_{\phi,v} : \mathbb{C} \smallsetminus \mathcal{J}^{\mathrm{f}}(\phi) \longrightarrow \mathbb{R}$$

is a harmonic function ($\Delta g_{\phi,v} = 0$) that vanishes on $\mathcal{J}(\phi) = \partial\mathcal{J}^{\mathrm{f}}(\phi)$ and has a simple logarithmic singularity as $|z|_v \to \infty$.

(e) **(Local Decomposition of the Canonical Height).** *Let L/K be a finite extension and let $\alpha \in \mathbb{A}^1(L)$. Then*

$$\hat{h}_\phi(\alpha) = \sum_{v\in M_L} \frac{[L_v : \mathbb{Q}_v]}{[L : \mathbb{Q}]} \hat{g}_{\phi,v}(\alpha).$$

A similar theory is available for morphisms on $\mathbb{P}^N$. The simplest construction is to lift ϕ to a polynomial map

$$\Phi : \mathbb{A}^{N+1} \longrightarrow \mathbb{A}^{N+1}$$

and define

$$\widehat{G}_{\Phi,v}(P) = \lim_{n\to\infty} \frac{1}{d^n} \log\bigl\|\Phi^n(P)\bigr\|_v,$$

where for an $(N+1)$-tuple $x = (x_0, \ldots, x_N)$, we let $\|x\|_v = \max_i\bigl\{|x_i|_v\bigr\}$. The function $\widehat{G}_{\Phi,v}$ satisfies the canonical transformation property

$$\widehat{G}_{\Phi,v} \circ \phi = d\widehat{G}_{\Phi,v},$$

but it does not give a well-defined function on $\mathbb{P}^N(\overline{K}_v)$ due to the ambiguity of lifting a point from $\mathbb{P}^N$ to $\mathbb{A}^{N+1}$. So instead one often works with the modified function

$$\hat{g}_{\phi,v} \colon \mathbb{P}^N(\overline{K}_v) \longrightarrow \mathbb{R}, \quad P \longmapsto \widehat{G}_{\Phi,v}(\bar{P}) - \log|\bar{P}|_v,$$

where $\bar{P}$ is any lift of P to $\mathbb{A}^{N+1}(\overline{K})$. For details, see [102] for $\mathbb{C}$ and [60] for the non-Archimedean case. In particular, local canonical heights are Hölder continuous, i.e., they satisfy

$$\bigl|\hat{g}_{\phi,v}(P) - \hat{g}_{\phi,v}(Q)\bigr| \ll \operatorname{dist}_v(P,Q),$$

where dist_v is the chordal distance on $\mathbb{P}^N(\overline{K}_v)$.

We also note that in the non-Archimedean setting, there has been extensive work extending the theory of Green functions to Berkovich space and applying the results to dynamics; see [4].

Finally, we consider the practical problem of computing the canonical height $\hat{h}_\phi(P)$ for specific maps and points. It is problematic to directly use the limit definition of $\hat{h}_\phi(P)$ because of the number of bits required to explicitly write down the value of $\phi^n(P)$ when n is large. It is often easier to compute and sum the individual local height functions. See [26] for a rapidly converging series for $\hat{g}_{\phi,v}$. (This series generalizes earlier work of Tate and Silverman on elliptic curves [104]. See also [108] for methods that help avoid having to factor large numbers.)

5.5. Specialization theorems

Consider a family of morphisms parametrized by a (smooth quasi-projective) variety T,

$$\phi \colon \mathbb{P}^N_T \longrightarrow \mathbb{P}^N_T.$$

In other words, we have a commutative diagram

$$\begin{array}{ccc} \mathbb{P}^N \times T & \xrightarrow{\quad\phi\quad} & \mathbb{P}^N \times T \\ & {\scriptstyle \mathrm{proj}_2}\searrow \quad \swarrow{\scriptstyle \mathrm{proj}_2} & \\ & T. & \end{array}$$

The map ϕ induces a morphism

$$T \longrightarrow \mathsf{M}^N_d, \quad t \longmapsto \langle \phi_t \rangle,$$

and it is natural to ask how the height $\hat{h}_{\phi_t}(x)$ varies with $x \in \mathbb{P}^N$ and $t \in T$.

Theorem 5.20. *Fix some projective embedding of T and use it to define a height h_T on T. Also let $h_{\mathbb{P}^N}$ be the usual height on $\mathbb{P}^N$. Then*

$$\hat{h}_{\phi_t}(x) = h_{\mathbb{P}^N}(x) + O_\phi\bigl(h_T(t)\bigr) \quad \textit{for all } x \in \mathbb{P}^N(\overline{K}) \textit{ and } t \in T(\overline{K}).$$

The O-constant depends on the family of maps ϕ and the choice of height h_T, but is independent of both x and t.

Proof Sketch. Fixing a projective completion $\overline{T}$ of T, we obtain a rational map

$$\bar{\phi} \colon \mathbb{P}^N_{\overline{T}} \dashrightarrow \mathbb{P}^N_{\overline{T}}.$$

We resolve the locus of indeterminacy of this rational map to get a (smooth projective) variety X with morphisms π and ψ fitting into the following commutative diagram:

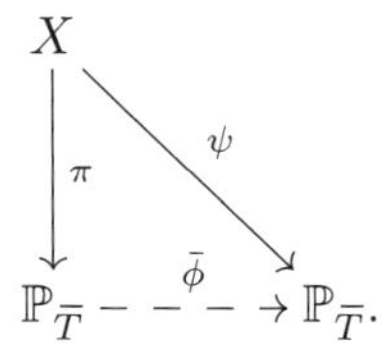

The rest of the proof is a computation using functorial properties of height functions applied to the morphisms π and ψ, plus some basic facts about ample divisors. For details, see [26]. $\square$

We now fix the following quantities:

K a number field,

T/K a smooth projective curve,

$\phi \in \operatorname{Hom}_d^N\big(K(T)\big)$, a family of maps.

$P \in \mathbb{P}^N\big(K(T)\big)$, a family of points.

There are three natural height functions to consider:

h_T We embed $T \subset \mathbb{P}^L$ and let $h_T = \dfrac{1}{\deg T} h_{\mathbb{P}^L}$.

$\hat{h}_\phi \colon \mathbb{P}^N\big(K(T)\big) \to \mathbb{R}$, the function field canonical height.

$\hat{h}_{\phi_t} \colon \mathbb{P}^N(\overline{K}) \to \mathbb{R}$, for $t \in T(\overline{K})$, the number field canonical height.

Theorem 5.21. *With notation as above,*

$$\lim_{\substack{t \in T(\overline{K}) \\ h_T(t) \to \infty}} \frac{\hat{h}_{\phi_t}(P_t)}{h_T(t)} = \hat{h}_\phi(P).$$

The next corollary follows immediately from the theorem.

Corollary 5.22. *If* $\hat{h}_\phi(P) \neq 0$, *then*

$$\big\{ t \in T(\overline{K}) : P_t \in \operatorname{PrePer}(\phi_t) \big\}$$

is a set of bounded height.

Proof Sketch of Theorem 5.21. A point $P \in \mathbb{P}^N\big(K(T)\big)$ is equivalent to a rational map

$$P \colon T \dashrightarrow \mathbb{P}^N,$$

but in our case T is a smooth curve and $\mathbb{P}^N$ is complete, so P is a morphism. An easy calculation shows that

$$\deg P^* H = h(P), \tag{5.7}$$

where $H \in \operatorname{Div}(\mathbb{P}^N)$ is a hyperplane. (To see this, note that $\deg P^* H = P(T) \cdot H$, and the intersection index is computed as a sum of local terms.)

We compute

$$\begin{aligned}
\hat{h}_{\phi_t}(P_t) &= h_{\mathbb{P}^N}(P_t) + O_\phi\big(h_T(t)\big) \\
&\qquad\qquad \text{from Theorem 5.20 with } x = P_t, \\
&= h_{T,P^*H}(t) + O_P(1) + O_\phi\big(h_T(t)\big) \\
&\qquad\qquad \text{functorial property of heights}, \\
&= (\deg P^*H)h_T(t) + O_P(1) + O_\phi\big(h_T(t)\big) \\
&\qquad\qquad \text{true for } T = \mathbb{P}^1, \text{ else error of } O_P\big(\sqrt{h_T(t)}\big), \\
&= h(P)h_T(t) + O_P(1) + O_\phi\big(h_T(t)\big) \\
&\qquad\qquad \text{from (5.7) we have } \deg P^*H = h(P), \\
&= \big(\hat{h}_\phi(P) + O_\phi(1)\big)h_T(t) + O_P(1) + O_\phi\big(h_T(t)\big) \\
&\qquad\qquad \text{since Theorem 5.5 says } \hat{h}_\phi = h + O(1).
\end{aligned}$$

Now divide by $h_T(t)$ and take the lim sup as $h_T(t) \to \infty$,

$$\limsup_{\substack{t\in T(\bar{K}) \\ h_T(t)\to\infty}} \left| \frac{\hat{h}_{\phi_t}(P_t)}{h_T(t)} - \hat{h}_\phi(P) \right| = O_\phi(1).$$

N.B. The key here is that the big-O constant is independent of P, since the $O_P(1)$ term was divided by $h_T(t)$ and then $h_T(t) \to \infty$.

We now replace P by $\phi^n(P)$ and use the fact that

$$\hat{h}_{\phi_t}(\phi^n P_t) = d^n \hat{h}_{\phi_t}(P_t) \quad \text{and} \quad \hat{h}_\phi(\phi^n P) = d^n \hat{h}_\phi(P)$$

to get

$$\limsup_{\substack{t\in T(\bar{K}) \\ h_T(t)\to\infty}} \left| \frac{\hat{h}_{\phi_t}(P_t)}{h_T(t)} - \hat{h}_\phi(P) \right| = \frac{1}{d^n} O_\phi(1).$$

The left-hand side is independent of n, so letting $n \to \infty$ gives the desired result. □

History. Theorems 5.20 and 5.21 were originally proven for abelian varieties, the former by Silverman and Tate, the latter by Silverman [103] for abelian varieties and by Tate [117] in strengthened form for elliptic curves. The proofs were adapted to the dynamical setting by Call and Silverman in [26], which also includes specialization theorems for local heights.

The estimate given by Theorem 5.21 can be strengthened for polynomial maps; cf. [45, 117] for analogous results for abelian varieties.

Theorem 5.23 (Ingram [54]). *Continuing with notation from Theorem 5.21, assume that $\phi(z) \in K(T)[z]$ is a polynomial. Then there is a divisor $D \in \mathrm{Div}(T)\otimes\mathbb{Q}$ such that*

$$\hat{h}_{\phi_t}(P_t) = h_{T,D}(t) + O(1) \quad \textit{for all } t \in T(\bar{K}).$$

In particular, $\deg D = \hat{h}_\phi(P)$, *and*

$$\hat{h}_{\phi_t}(P_t) = \hat{h}_\phi(P)h_T(t) + \begin{cases} O(1) & \textit{if } \mathrm{genus}(T) = 0, \\ O\big(\sqrt{h_T(t)}\big) & \textit{if } \mathrm{genus}(T) \ge 1. \end{cases}$$

Ingram [55] has also proven a version of Theorem 5.23 for families of Hénon maps.

Theorem 5.21 says that if the function field canonical height $\hat{h}_\phi(P)$ is not zero, then the set of $t \in T(\overline{K})$ with P_t preperiodic is a set of bounded height. This implies that there are only finitely many such t defined over any given number field, but in general there will be infinitely many such t defined over $\overline{K}$. If we impose additional conditions on t, we obtain the following dynamical analogue of a conjecture on "unlikely intersections" for algebraic groups due in various forms to Bombieri, Masser, Pink, Zannier, and Zilber [18, 20, 98, 124]; see also the work on Habegger [46, 47] in which he proves unlikely intersection theorems for tori and for abelian varieties.

Conjecture 5.24 (Unlikely Intersection Conjecture). *Let $T/\mathbb{C}$ be a smooth projective curve with function field $\mathbb{C}(T)$, and let $\phi \in \operatorname{Hom}_d^N(\mathbb{C}(T))$ be a family of maps. We assume that ϕ is not isotrivial (see Section 3.1). Let $P, Q \in \mathbb{P}^N(\mathbb{C}(T))$ be families of points that are* independent *in the sense that*

$$\#\mathcal{O}_\phi(P) = \infty, \quad \#\mathcal{O}_\phi(Q) = \infty, \quad \textit{and} \quad \mathcal{O}_\phi(P) \cap \mathcal{O}_\phi(Q) = \varnothing.$$

(Note that these are orbits over the function field $\mathbb{C}(T)$.) Then the set

$$\bigl\{t \in \mathbb{P}^1(\mathbb{C}) : P_t \textit{ and } Q_t \textit{ are both in } \operatorname{PrePer}(\phi_t)\bigr\}$$

is a finite set.

The intuition for Conjecture 5.24 rests on the following two ideas:

- Assuming that we work over $\overline{\mathbb{Q}}(T)$, rather than $\mathbb{C}(T)$, Corollary 5.22 says that requiring either of P_t or Q_t to be preperiodic forces t to lie in a set of bounded height.
- Two "independent" sets of bounded height should have finite intersection.

Here is a special case of the conjecture whose proof is far from elementary.

Theorem 5.25 (Baker–DeMarco [3]). *Fix $d \ge 2$ and $a, b \in \mathbb{C}$, and assume that $a^d \ne b^d$. Then*

$$\bigl\{c \in \mathbb{C} : a, b \in \operatorname{PrePer}(z^d + c)\bigr\} \quad \textit{is a finite set.}$$

Remark 5.26. Ghioca, Hsia, and Tucker [43] have proven a more general version of Theorem 5.25 in which the polynomials and the points are allowed to simultaneously vary in algebraic families, subject to a technical condition.

Remark 5.27. It would be interesting to formulate and prove versions of Corollary 5.22 and Conjecture 5.24 in which the parameter space T has dimension greater than one. Roughly speaking, if $\dim(T) = e$, then requiring that e "independent" points be preperiodic should give a set of bounded height in $T(\overline{\mathbb{Q}})$, and requiring that $e + 1$ "independent" points be preperiodic should give a finite subset of $T(\mathbb{C})$. However, even with the correct definition of independence, there are likely to be additional subtleties in defining the exceptional locus of T. See [19, 20, 46, 47, 71] for a discussion of specialization for higher dimensional families of tori and abelian varieties.

Let $\phi \in \operatorname{Hom}_d^N(\overline{\mathbb{Q}})$ and $\psi \in \operatorname{Hom}_e^N(\overline{\mathbb{Q}})$ be self-maps of $\mathbb{P}^N$ of degree at least 2. For each $P \in \mathbb{P}^N(\overline{\mathbb{Q}})$, we consider $\mathcal{O}_{\phi,\psi}(P)$, the full forward orbit of P under the semigroup of maps generated by ϕ and ψ. If the iterates of ϕ and ψ commute, then using properties of the canonical height it is easy to see that for all points $P \in \mathbb{P}^N(\overline{\mathbb{Q}})$ we have

$$\#\{Q \in \mathcal{O}_{\phi,\psi}(P) : h(Q) \le B\} \ll (\log B)^2.$$

On the other hand, if the iterates of ϕ and ψ do not commute, then we would expect that for "most" points $P \in \mathbb{P}^N(\overline{\mathbb{Q}})$, the full forward orbit $\mathcal{O}_{\phi,\psi}(P)$ should be "large." The following question quantifies the idea.

Question 5.28. Let $\phi \in \operatorname{Hom}_d^N(\overline{\mathbb{Q}})$ and $\psi \in \operatorname{Hom}_e^N(\overline{\mathbb{Q}})$ be self-maps of $\mathbb{P}^N$ of degree at least 2. Suppose that for every $\epsilon > 0$ and every point $P \in \mathbb{P}^N(\overline{\mathbb{Q}})$ we have

$$\#\{Q \in \mathcal{O}_{\phi,\psi}(P) : h(Q) \le B\} \ll B^\epsilon \quad \text{as } B \to \infty,$$

where the implied constant may depend on ϕ, ψ, and P. Does it follow that some iterate of ϕ commutes with some iterate of ψ? How about if we replace the upper bound with a power of $\log(B)$? One can obviously ask the same question for the orbit generated by more than two maps.

5.6. Heights and dominant rational maps

We have not talked a lot about rational maps $\phi\colon \mathbb{P}^N \dashrightarrow \mathbb{P}^N$ for several reasons. First, it may not make sense to talk about iteration, since for example the image $\phi(\mathbb{P}^N)$ of ϕ might lie in the indeterminacy locus $Z(\phi)$ of ϕ.[1] However, if we insist that ϕ be dominant, i.e., $\phi(\mathbb{P}^N) = \mathbb{P}^N$, then the iterates of ϕ are well-defined rational maps of $\mathbb{P}^N$. Second, even if all of the iterates of ϕ are well-defined, the orbit of a point may not be well-defined because some iterate $\phi^n(P)$ might be in $Z(\phi)$, and then $\phi^{n+1}(P)$ is not defined.

Example 5.29. An important situation in which iteration and orbits are well-defined is the case of an extension of a dominant morphism $\mathbb{A}^N \to \mathbb{A}^N$ to a rational map $\mathbb{P}^N \dashrightarrow \mathbb{P}^N$. The dynamics of ϕ on $\mathbb{A}^N$ are then quite interesting. See Section 5.7 for a discussion of the case that the map $\mathbb{A}^N \to \mathbb{A}^N$ is a (regular) automorphism.

Proposition 5.4 tells us that if $\phi : \mathbb{P}^N \to \mathbb{P}^N$ is a morphism of degree d, then it transforms the height in a consistent manner,

$$h\big(\phi(P)\big) = dh(P) + O(1) \quad \text{for all } P \in \mathbb{P}^N(\overline{\mathbb{Q}}).$$

If ϕ is simply a rational map of degree d, then we still have an elementary upper bound $h\big(\phi(P)\big) \le dh(P) + O(1)$, but the lower bound no long holds. However, if ϕ is dominant, there does exist a weaker lower bound.

Proposition 5.30. *Let $\phi\colon \mathbb{P}^N \dashrightarrow \mathbb{P}^N$ be a dominant rational map. Then there are a nonempty Zariski open set $U \subset \mathbb{P}^N$ and constants $c_1 > 0$ and c_2 such that*

$$h\big(\phi(P)\big) \ge c_1 h(P) - c_2 \qquad \textit{for all } P \in U(\overline{\mathbb{Q}}).$$

PROOF. See [112] for two proofs of a stronger statement, one using facts from algebraic geometry, the other an elementary argument using basic properties of heights. □

[1]For a rational map $\phi\colon X \dashrightarrow Y$ between smooth varieties, we define the image $\phi(X)$ by taking a nonempty Zariski open set $U \subset X$ on which ϕ is defined and setting $\phi(X)$ to be the Zariski closure of $\phi(U)$ in Y.

Definition 5.31. Proposition 5.30 says that the ratio $h(\phi(P))/h(P)$ is bounded below on a large subset of $\mathbb{P}^N(\mathbb{Q})$. The largest such bound is called the *height expansion coefficient of* ϕ and is defined by

$$\mu(\phi) = \limsup_{\varnothing \ne U \subset \mathbb{P}^N \smallsetminus Z(\phi)} \liminf_{P \in U(\overline{\mathbb{Q}})} \frac{h(\phi(P))}{h(P)}.$$

The supremum is over all nonempty Zariski open subsets of $\mathbb{P}^N$ on which ϕ is defined.

Remark 5.32. We note that

$$\mu(\phi) = \mu(\phi^f) \quad \text{for all } f \in \mathrm{PGL}_{N+1}.$$

This follows easily from the fact that $f\colon \mathbb{P}^N \to \mathbb{P}^N$ and its inverse are morphisms of degree 1, so they only change the height by an $O(1)$, which disappears when we divide by $h(P)$ and let $h(P) \to \infty$.

Example 5.33. Let $\phi\colon \mathbb{P}^N \dashrightarrow \mathbb{P}^N$ be a regular affine automorphism (see Definition 2.59) of degree d whose indeterminacy locus $Z(\phi)$ has dimension 0. Then

$$\mu(\phi) = \frac{1}{d^{N-1}}.$$

(See [112, Proposition 10].)

Example 5.34. Let $\phi\colon \mathbb{P}^N \dashrightarrow \mathbb{P}^N$ be the dominant rational map of degree N defined by

$$\phi([X_0,\ldots,X_N]) = [X_0^{-1},\ldots,X_N^{-1}].$$

Then

$$\mu(\phi) = \frac{1}{N}.$$

(See [112, Proposition 11].)

If $\phi\colon \mathbb{P}^N \to \mathbb{P}^N$ is a morphism of degree d, then $\mu(\phi) = d$. Conversely, if ϕ is not a morphism, then ChongGyu Lee has shown that $\mu(\phi) \le d-1$; see [112, Proposition 9]. This suggests asking how small $\mu(\phi)$ can be.

Proposition 5.30 implies that $\mu(\phi) > 0$. Examples 5.33 and 5.34 show that $\mu(\phi)$ may be arbitrarily small if we take d and/or N to be large, but it is not clear what happens if we fix d and N and let ϕ vary.

Theorem 5.35. *The quantity*

$$\overline{\mu}_d(\mathbb{P}^N) = \inf_{\substack{\phi\colon \mathbb{P}^N \dashrightarrow \mathbb{P}^N \\ \phi \text{ dominant}}} \mu(\phi)$$

is strictly positive.

Proof. The proof uses a generalized version of Proposition 5.30 applied to the universal family of dominant rational self-maps of $\mathbb{P}^N$. A key part of the proof requires the fact that

$$\{\phi \in \mathrm{Hom}_d^N : \phi \text{ is dominant}\}$$

is a Zariski open subset of Hom_d^N. See [112, Section 3] for details. □

Remark 5.36. Example 5.33 shows that if $N \geq 2$, then

$$\overline{\mu}_d(\mathbb{P}^N) \leq d^{-(N-1)}.$$

It seems to be an interesting question to determine the exact value of $\overline{\mu}_d(\mathbb{P}^N)$, other than the trivial case $\overline{\mu}_d(\mathbb{P}^1) = d$. Another question raised in [112] is whether the quantity

$$\frac{\log \mu(\phi)}{\log \deg \phi}$$

is a rational number for all dominant rational maps $\phi\colon \mathbb{P}^N \dashrightarrow \mathbb{P}^N$.

5.7. Canonical heights for regular affine automorphisms

Recall from Definition 2.59 that a rational map $\phi\colon \mathbb{P}^N \dashrightarrow \mathbb{P}^N$ is a *regular affine automorphism* if:

(1) ϕ induces an automorphism $\mathbb{A}^n \to \mathbb{A}^n$.
(2) The loci of indeterminacy of ϕ and ϕ^{-1} are disjoint.

Theorem 5.37. *Let $\phi\colon \mathbb{P}^N \dashrightarrow \mathbb{P}^N$ be a regular affine automorphism defined over $\overline{\mathbb{Q}}$, and let*

$$d_1 = \deg(\phi) \quad \text{and} \quad d_2 = \deg(\phi^{-1}).$$

Then

$$\frac{1}{d_1}h\bigl(\phi(P)\bigr) + \frac{1}{d_2}h\bigl(\phi^{-1}(P)\bigr) \geq \left(1 + \frac{1}{d_1 d_2}\right) h(P) \quad \text{for all } P \in \mathbb{P}^N(\overline{\mathbb{Q}}).$$

PROOF. Theorem 5.37 was proven by Silverman [106] for Hénon maps of degree 2 on $\mathbb{P}^2$, then Kawaguchi [57] proved it for all maps on $\mathbb{P}^2$, and then Kawaguchi [58] and ChongGyu Lee [64, 65] proved it independently (by different methods) on $\mathbb{P}^N$. □

Using Theorem 5.37 (in a stronger local formulation), Kawaguchi [58] constructs canonical heights for regular affine automorphisms.

Theorem 5.38 (Kawaguchi [58]). *Let $\phi\colon \mathbb{P}^N \dashrightarrow \mathbb{P}^N$ be a regular affine automorphism defined over $\overline{\mathbb{Q}}$, and let*

$$d_1 = \deg(\phi) \quad \text{and} \quad d_2 = \deg(\phi^{-1}).$$

Then the limits

$$\hat{h}_\phi^+(P) = \lim_{n\to\infty} \frac{1}{d_1^n} h\bigl(\phi^n(P)\bigr) \quad \text{and} \quad \hat{h}_\phi^-(P) = \lim_{n\to\infty} \frac{1}{d_2^n} h\bigl(\phi^{-n}(P)\bigr)$$

exist. Define

$$\hat{h}_\phi(P) = \hat{h}_\phi^+(P) + \hat{h}_\phi^-(P).$$

Then $\hat{h}_\phi$, $\hat{h}_\phi^+$, and $\hat{h}_\phi^-$ have the following properties:

(a) $\hat{h}_\phi^+\bigl(\phi(P)\bigr) = d_1 \hat{h}_\phi^+(P)$ *and* $\hat{h}_\phi^-\bigl(\phi^{-1}(P)\bigr) = d_2 \hat{h}_\phi^-(P)$.

(b) $\frac{1}{d_1}\hat{h}_\phi\bigl(\phi(P)\bigr) + \frac{1}{d_2}\hat{h}_\phi\bigl(\phi^{-1}(P)\bigr) \geq \left(1 + \frac{1}{d_1 d_2}\right)\hat{h}_\phi(P)$.

(c) $h(P) + O(1) \leq \hat{h}_\phi(P) \leq 2h(P) + O(1)$.

(d) $\hat{h}_\phi^+(P) \geq 0$ *and* $\hat{h}_\phi^-(P) \geq 0$ *and* $\hat{h}_\phi(P) \geq 0$.

(e) $\hat{h}_\phi^+(P) = 0 \iff \hat{h}_\phi^-(P) = 0 \iff \hat{h}_\phi(P) = 0 \iff P \in \operatorname{Per}(\phi)$.

Kawaguchi's canonical height theorem immediately gives a Northcott-type theorem for regular affine automorphisms. (Note that for an automorphism, every preperiodic point is periodic.)

Corollary 5.39. *Let $\phi\colon \mathbb{P}^N \dashrightarrow \mathbb{P}^N$ be a regular affine automorphism defined over $\overline{\mathbb{Q}}$. Then*

$$\operatorname{Per}\bigl(\phi, \mathbb{A}^N(\overline{\mathbb{Q}})\bigr)$$

is a set of bounded height.

PROOF. The corollary is an immediate consequence of Theorem 5.38, but we note that it was originally proven by Denis [31] and Marcello [69, 70] using a weaker version of Theorem 5.37 that lacked the $1/d_1d_2$ in the lower bound. □

In view of Corollary 5.39, it is tempting to make a uniform boundedness conjecture for regular affine automorphisms.

Conjecture 5.40 (Uniform Boundededness for Regular Affine Automorphisms). *For all N, d, D there is a constant $C(N, d, D)$ such that for all number fields $K/\mathbb{Q}$ of degree at most D and all regular affine automorphisms $\phi\colon \mathbb{P}^N \dashrightarrow \mathbb{P}^N$ of degree d defined over K,*

$$\#\operatorname{Per}\bigl(\phi, \mathbb{P}^N(K)\bigr) \le C(N, d, D).$$

Ingram has made a precise conjecture for certain Hénon maps.

Conjecture 5.41 (Ingram [55]). *For $b \in \mathbb{Q}$, let ϕ_b be the Hénon map*

$$\phi_b(x, y) = (x, x + y^2 + b).$$

If $P \in \mathbb{A}^2(\mathbb{Q})$ has exact period n for ϕ_b, then $n \in \{1, 2, 3, 4, 6, 8\}$.

Ingram has verified that Conjecture 5.41 is true for all rational numbers $b = A/B$ with $\max\bigl\{|A|, |B|\bigr\} \le 1000$. He also cites the map $\phi_{-9/16}(x, y)$ as having several rational periodic points, including two points $(\frac{3}{4}, \frac{3}{4})$ and $(-\frac{3}{4}, -\frac{3}{4})$ that are fixed, a point $(\frac{3}{4}, -\frac{3}{4})$ of period 2, and a point $(\frac{1}{4}, -\frac{3}{4})$ of period 8.

5.8. Canonical heights for K3 dynamics

We continue with the K3 surfaces

$$S_{\boldsymbol{A},\boldsymbol{B}} \subset \mathbb{P}^2 \times \mathbb{P}^2$$

described in Section 3.3. For notational convenience, we drop the subscripts and write S for this surface. For details of the material covered in this section, see [105] and [110, §7.4].

Let $H \subset \mathbb{P}^2$ be a line. The Picard group of $\mathbb{P}^2 \times \mathbb{P}^2$ is generated by the divisors

$$H_1 = H \times \mathbb{P}^2 \quad \text{and} \quad H_2 = \mathbb{P}^2 \times H,$$

and the restriction of these divisors to S gives divisors

$$D_1 = H_1\big|_S \quad \text{and} \quad D_2 = H_2\big|_S. \tag{5.8}$$

Recall that there are involutions $\iota_1, \iota_2\colon S \to S$ associated to the two double covers $S \to \mathbb{P}^2$ induced by the two projection maps $\mathbb{P}^2 \times \mathbb{P}^2 \to \mathbb{P}^2$. An elementary calculation gives linear equivalences

$$\iota_1^* D_1 \sim D_1, \quad \iota_2^* D_1 \sim -D_1 + 4D_2, \quad \iota_2^* D_1 \sim 4D_1 - D_2, \quad \iota_2^* D_2 \sim D_2. \tag{5.9}$$

Setting

$$\alpha = 2 + \sqrt{3} \qquad \text{and} \qquad \phi = \iota_1 \circ \iota_2 \in \operatorname{Aut}(S),$$

it follows from (5.9) that the divisors

$$E_1 = -D_1 + \alpha D_2 \quad \text{and} \quad E_2 = \alpha D_1 - D_2 \tag{5.10}$$

in $\operatorname{Div}(S) \otimes \mathbb{R}$ satisfy

$$\phi^* E_1 \sim \alpha^2 E_1 \quad \text{and} \quad (\phi^{-1})^* E_2 \sim \alpha^2 E_2. \tag{5.11}$$

In order to study arithmetic properties of iteration by ϕ, we need to translate the geometric information encoded in (5.11) into arithmetic information. This is done using height functions.

In general, if V is a projective variety, each embedding of V into projective space defines a height function on the points of V. Weil's height machine assigns a height function to each divisor on V.

Theorem 5.42 (Weil's Height Machine). *Let K be a global field, i.e., a number field or a 1-dimensional function field. For every smooth projective variety V/K there exists a map*

$$h_V\colon \operatorname{Div}(V) \longrightarrow \{\text{functions } V(\overline{K}) \to \mathbb{R}\}$$

with the following properties:

(a) **(Normalization)** *Let $H \subset \mathbb{P}^N$ be a hyperplane and let $h\colon \mathbb{P}^N(\overline{K}) \to \mathbb{R}$ be the absolute logarithmic height on $\mathbb{P}^N$. Then*

$$h_{\mathbb{P}^N,H}(P) = h(P) + O(1) \qquad \text{for all } P \in \mathbb{P}^N(\overline{K}).$$

(b) **(Functoriality)** *Let $\phi\colon V \to W$ be a morphism and let $D \in \operatorname{Div}(W)$. Then*

$$h_{V,\phi^* D}(P) = h_{W,D}\big(\phi(P)\big) + O(1) \quad \text{for all } P \in V(\overline{K}).$$

(c) **(Additivity)** *Let $D, E \in \operatorname{Div}(V)$. Then*

$$h_{V,D+E}(P) = h_{V,D}(P) + h_{V,E}(P) + O(1) \quad \text{for all } P \in V(\overline{K}).$$

(d) **(Linear Equivalence)** *Let $D, E \in \operatorname{Div}(V)$ with D linearly equivalent to E. Then*

$$h_{V,D}(P) = h_{V,E}(P) + O(1) \quad \text{for all } P \in V(\overline{K}).$$

(e) **(Finiteness)** *Let $D \in \operatorname{Div}(V)$ be ample. If K is a number field, then the set*

$$\big\{P \in V(\overline{K}) : h_{V,D}(P) \le A \text{ and } [K(P) : K] \le B\big\}$$

is finite, and if K is a function field, then it is a finite union of finite dimensional families, cf. Proposition 5.2.

(f) **(Uniqueness)** *The height functions $h_{V,D}$ are determined, up to $O(1)$, by normalization* (a), *functoriality* (b) *for embeddings $\phi\colon V \hookrightarrow \mathbb{P}^n$, and additivity* (c).

PROOF. See [17,48,62] for an exposition of this classical theory, as well as some additional useful properties of the height machine. □

Definition 5.43. A subset $W \subset V(\overline{K})$ is said to be a *set of bounded height* if there is an ample divisor $D \in \operatorname{Div}(V)$ such that

$$\sup_{P\in W} h_{V,D}(P) \tag{5.12}$$

is bounded. One can show that if (5.12) is bounded for one ample divisor D, then it bounded for all ample divisors. If K is a number field and $W \subset V(\overline{K})$ is a set of bounded height, then Theorem 5.42(e) implies that

$$W \cap \bigl\{P \in V(\overline{K}) : [K(P):K] \le B\bigr\}$$

is finite.

Remark 5.44. Using the additivity property, we can extend h_V to a map

$$h_V\colon \operatorname{Div}(V)\otimes\mathbb{R} \longrightarrow \{\text{functions } V(\overline{K}) \to \mathbb{R}\},$$

$$\sum \alpha_i D_i \longmapsto \sum \alpha_i h_{V,D_i}.$$

Applying the height machine and Tate's telescoping sum trick gives canonical heights on the K3 surface S relative to the map ϕ and the divisors E_1 and E_2.

Theorem 5.45. *Let $S \subset \mathbb{P}^2 \times \mathbb{P}^2$ be a nonsingular* K3 *surface defined by the intersection of a $(2,2)$-hypersurface and a $(1,1)$-hypersurface, let $D_1, D_2, E_1, E_2 \in \operatorname{Div}(S)\otimes\mathbb{R}$ be as defined by* (5.8) *and* (5.10), *and let $\phi = \iota_1\circ\iota_2\colon S \to S$. There are canonical height functions*

$$\hat h_\phi^+\colon S(\overline{K}) \to \mathbb{R} \quad \text{and} \quad \hat h_\phi^-\colon S(\overline{K}) \to \mathbb{R}$$

with the following properties: For all $P \in S(\overline{K})$,

(a) $\hat h_\phi^+(P) = h_{S,E_1}(P) + O(1)$ *and* $\hat h_\phi^-(P) = h_{S,E_2}(P) + O(1)$.

(b) $\hat h_\phi^+(\phi(P)) = \alpha^2 \hat h_\phi^+(P)$ *and* $\hat h_\phi^-(\phi(P)) = \alpha^{-2}\hat h_\phi^+(P)$.

(c) *Let*

$$\hat h_\phi = \hat h_\phi^+ + \hat h_\phi^-,$$

and assume that K is a number field. Then the following are equivalent:

(i) $\hat h_\phi^+(P) = 0$.

(ii) $\hat h_\phi^-(P) = 0$.

(iii) $\hat h_\phi(P) = 0$.

(iv) *P is ϕ-periodic.*

Further, for all A and B the following set is finite:

$$\bigl\{P \in S(\overline{K}) : \hat h_\phi(P) \le A \text{ and } [K(P):K] \le B\bigr\} \tag{5.13}$$

Proof Sketch. Applying Theorem 5.42 to the linear equivalence

$$\phi^* E_1 \sim \alpha^2 E_1$$

from (5.11) gives

$$h_{S,E_1}\circ\phi = h_{S,\phi^*E_1} + O(1) = h_{S,\alpha^2E_1} + O(1) = \alpha^2 h_{S,E_1} + O(1),$$

and similarly $h_{S,E_2}\circ\phi^{-1} = \alpha^2 h_{S,E_2} + O(1)$. Then a telescoping sum argument, cf. the proof of Theorem 5.5, shows that the limits

$$\begin{aligned} \hat h_\phi^+(P) &= \lim_{n\to\infty} \alpha^{-2n} h_{S,E_1}\bigl(\phi^n(P)\bigr),\\ \hat h_\phi^-(P) &= \lim_{n\to\infty} \alpha^{-2n} h_{S,E_2}\bigl(\phi^{-n}(P)\bigr), \end{aligned} \tag{5.14}$$

exist and satisfy (a). These limit formulas immediately give (b).

For (c), we note that $E_1 + E_2 = (\alpha - 1)(D_1 + D_2)$ is in the ample cone and that

$$\hat{h}_\phi = h_{S,E_1+E_2} + O(1) = (\alpha - 1)h_{S,D_1+D_2} + O(1),$$

so $S(\overline{K})$ contains only finitely many points having bounded $\hat{h}_\phi$-height and bounded degree over K. This gives (5.13). Then

$$\begin{aligned} \hat{h}_\phi(P) = 0 &\iff \hat{h}_\phi\bigl(\phi^n(P)\bigr) = 0 \\ &\iff h_{S,D_1+D_2}\bigl(\phi^n(P)\bigr) = O(1) \\ &\iff \mathcal{O}_\phi(P) \text{ has bounded height} \\ &\iff \mathcal{O}_\phi(P) \text{ is finite} \\ &\iff P \text{ is } \phi\text{-periodic.} \end{aligned}$$

Next, since $\hat{h}_\phi = \hat{h}_\phi^+ + \hat{h}_\phi^-$ and all three heights are nonnegative, we see that

$$\hat{h}_\phi(P) = 0 \iff \hat{h}_\phi^+(P) = 0 \quad \text{and} \quad \hat{h}_\phi^-(P) = 0.$$

Finally, suppose that $\hat{h}_\phi^+(P) = 0$. Then

$$\begin{aligned} \hat{h}_\phi\bigl(\phi^n(P)\bigr) &= \hat{h}_\phi^+\bigl(\phi^n(P)\bigr) + \hat{h}_\phi^-\bigl(\phi^n(P)\bigr) \\ &= \alpha^{2n}\hat{h}_\phi^+(P) + \alpha^{-2n}\hat{h}_\phi^-(P) \\ &= \alpha^{-2n}\hat{h}_\phi^-(P). \end{aligned}$$

The right-hand side is bounded as $n \to \infty$. Hence $\{\phi^n(P) : n \geq 0\}$ is a set of bounded height, which implies that it is a finite set, so P is periodic. Then the limit formula (5.14) shows that $\hat{h}_\phi^-(P) = 0$. This proves that

$$\hat{h}_\phi^+(P) = 0 \iff \hat{h}_\phi^-(P) = 0,$$

and the same argument, *mutatis mutandis*, gives the opposite implication. □

Corollary 5.46. *Let $S = S_{\boldsymbol{A},\boldsymbol{B}} \subset \mathbb{P}^2 \times \mathbb{P}^2$ be a nonsingular* K3 *surface and let $\phi = \iota_1 \circ \iota_2$. Then $\operatorname{Per}(S, \phi)$ is a set of bounded height. In particular, if K is a number field, then $S(K)$ contains only finitely many ϕ-periodic points.*

As in the case of regular affine automorphisms, anytime one has a Northcott-type finiteness theorem, it is tempting to make a uniform boundedness conjecture.

Conjecture 5.47 (Uniform Boundededness for K3 Surfaces). *For all N and D there is a constant $C(N, D)$ such that for all number fields $K/\mathbb{Q}$ of degree at most D and all nonsingular* K3 *surfaces $S_{\boldsymbol{A},\boldsymbol{B}} \subset \mathbb{P}^2 \times \mathbb{P}^2$ defined over K with automorphism $\phi = \iota_1 \circ \iota_2$,*

$$\#\operatorname{Per}\bigl(\phi, S_{\boldsymbol{A},\boldsymbol{B}}(K)\bigr) \leq C(N, D).$$

Example 5.48. One can also construct canonical heights on other K3 surfaces, for example the surfaces $S \subset \mathbb{P}^1 \times \mathbb{P}^1 \times \mathbb{P}^1$ described in Example 3.19. Continuing with the notation from that example, each automorphism such as $\phi = \iota_{12} \circ \iota_{13}$ and $\psi = \iota_{12} \circ \iota_{23}$ gives canonical heights relative to eigendivisors in $\operatorname{Div}(S) \otimes \mathbb{R}$, i.e., divisors E satisfying

$$\phi^* E \sim cE \quad \text{for some } c > 1.$$

(Of course, if $c < 1$, then we can use $(\phi^{-1})^*E \sim c^{-1}E$.) The automorphism group $\mathrm{Aut}(S)$ is large, since for example ϕ and ψ generate a free group, so there will be many different canonical heights associated to different maps and different divisors. Baragar has suggested using vector-valued heights. See [7, 9, 10, 121].

5.9. Algebraic dynamics and transcendental numbers

There are many quantities appearing in algebraic dynamics that are obviously composed of algebraic numbers. For example, if $\phi(z) \in \overline{\mathbb{Q}}(z)$, then the critical points of ϕ and the preperiodic points of ϕ are algebraic numbers. Similarly, it follows from the upcoming Theorem 6.7 in Section 6.1 that the points corresponding to PCF maps in the moduli space M_d are actually algebraic points, i.e., they lie in $\mathsf{M}_d(\overline{\mathbb{Q}})$. These facts are dynamical analogues of the facts from arithmetic geometry that the torsion points of an abelian variety $A/\overline{\mathbb{Q}}$ lie in $A(\overline{\mathbb{Q}})$ and that the points the moduli space A_g of principally polarized abelian varieties of dimension g corresponding to CM abelian varieties lie in $\mathsf{A}_g(\overline{\mathbb{Q}})$.

There are also many quantities in arithmetic geometry that are either known or conjectured to be transcendental over $\mathbb{Q}$. These numbers tend to be periods or values of suitably generalized logarithm functions. Indeed, the field of transcendental number theory has a long history and includes many deep results. This suggests looking for an analogous transcendence theory in dynamics.

Theorem 5.49 (Becker–Bergweiler [11, 12]; see also [120]). *Let $F(z)$, $G(z) \in \mathbb{C}[z]$ be polynomials of the same degree. A classical theorem of Böttcher says that there exists an analytic function $\xi(z)$ that is conformal in a neighborhood U of ∞ such that*

$$F\big(\xi(z)\big) = \xi\big(G(z)\big) \quad \text{for } z \in U,$$

i.e., the functions F and G are locally conjugate around ∞. Assume further that F and G are not linearly conjugate, i.e., not conjugate by a linear map $z \to az + b$, and that neither F nor G is linearly conjugate to z^d or to a Chebyshev polynomial. Then

$$\xi(\alpha) \text{ is transcendental for all } \alpha \in U \cap \overline{\mathbb{Q}}.$$

This theorem admits a nice corollary related to the Douady–Hubbard proof that the Mandelbrot set is connected.

Corollary 5.50. *Let*

$$\mathfrak{M} = \{c \in \mathbb{C} : 0 \text{ is bounded under iteration of } z^2 + c\}$$

denote the Mandelbrot set, and let

$$\psi\colon \mathbb{C} \smallsetminus \mathfrak{M} \longrightarrow \big\{z \in \mathbb{C} : |z| > 1\big\}$$

be the conformal map defined by Douady and Hubbard [34]. Then

$$\psi(\alpha) \text{ is transcendental for all } \alpha \in \overline{\mathbb{Q}} \smallsetminus \mathfrak{M}.$$

One might also ask about the values of dynamical canonical height functions, both local and global. In the setting of arithmetic geometry, it has been conjectured that if $E/\overline{\mathbb{Q}}$ is an elliptic curve and $P \in E(\overline{\mathbb{Q}})$ a nontorsion point, then $\hat{h}(P)$ and $\exp\big(\hat{h}(P)\big)$ are transcendental. This is not currently known for a single example, although Bertrand [14] has shown that certain p-adic valued canonical heights are transcendental.

Example 5.51. It is easy to see that the general dynamical analogue of this elliptic curve conjecture is false. For example, taking the map $\phi(z) = z^d$, we have $\hat{h}_\phi(\alpha) = h(\alpha)$, so in particular for every $a/b \in \mathbb{Q}$,

$$\exp\left(\hat{h}_\phi\left(\frac{a}{b}\right)\right) = \exp\left(\log\max\{|a|,|b|\}\right) = \max\{|a|,|b|\} \in \mathbb{Z}.$$

It is not hard to concoct similar examples for maps whose dynamics is more complicated. For example, let

$$\phi(z) = 1 - z^2 \quad \text{and} \quad \beta = \frac{1}{b} \in \mathbb{Q} \quad \text{with } b \geq 2.$$

Then one easily checks that

$$\phi^n(\beta) = \frac{A_n}{b^{2^n}} \quad \text{with } 0 < A_n < b^{2^n},$$

so

$$\hat{h}_\phi(\beta) = \lim_{n\to\infty} \frac{1}{2^n}\log\max\{A_n, b^{2^n}\} = \log(b).$$

Hence $\exp\bigl(\hat{h}_\phi(\beta)\bigr) = b \in \mathbb{Z}$.

These examples are easily explained. For $\phi(z) = z^d$, the point a/b is either preperiodic (equal to ± 1), or else it is in the attracting basin of 0 or ∞, which are superattracting fixed points of ϕ. For $\phi(z) = 1 - z^2$, the point $1/b$ is in the attracting basin of the superattracting two-cycle $\{0, 1\}$. More generally, it seems reasonable to suppose that $\exp\bigl(\hat{h}_\phi(P)\bigr)$ might sometimes take on algebraic values at algebraic points of the Fatou set, where iteration is well-behaved. This prompts the following conjecture which, via consideration of Lattès maps, generalizes the conjecture on elliptic curves.

Conjecture 5.52. *Let $\phi(z) \in \overline{\mathbb{Q}}(z)$ be a rational function of degree at least two whose Julia set is all of $\mathbb{P}^1(\mathbb{C})$. Then*

$$\hat{h}_f(\alpha) \textit{ and } \exp\bigl(\hat{h}_f(\alpha)\bigr) \textit{ are transcendental for all } \alpha \in \mathbb{P}^1(\overline{\mathbb{Q}}) \smallsetminus \mathrm{PrePer}(\phi).$$

Example 5.53. It is tempting to conjecture a similar result under the weaker assumption that one or more of the Galois conjugates of α lies in the Julia set of ϕ, but care must be taken. For example, consider the map $\phi(z) = z^d$. The Julia set of ϕ is the unit circle, $\mathcal{J}(\phi) = S^1 = \{z \in \mathbb{C} : |z| = 1\}$, and the preperiodic points for ϕ are the roots of unity, all of which lie in $\mathcal{J}(\phi)$. The number $(3+4i)/5$ and its conjugate $(3-4i)/5$ are both in $\mathcal{J}(\phi)$, while it is easy to compute $\exp\bigl(\hat{h}_\phi((3+4i)/5)\bigr) = 5$.

For this example, if we require that α be an algebraic integer and that all of its Galois conjugates are in $\mathcal{J}(\phi)$, then Kronecker's theorem implies that α is a root of unity. But suppose that we only impose the weaker condition that at least one of the Galois conjugates of α is in $\mathcal{J}(\phi)$. In that case, consider a Salem number α, which is a real algebraic number whose Galois conjugates

$$\alpha = \alpha_1, \alpha_2, \ldots, \alpha_d$$

have the following properties:

- $\alpha = \alpha_1 > 1$,
- $|\alpha_i| \leq 1$ for all $2 \leq i \leq d$,
- $|\alpha_i| = 1$ for some $2 \leq i \leq d$.

The smallest known Salem number is the largest real root of the polynomial

$$x^{10} + x^9 - x^7 - x^6 - x^5 - x^4 - x^3 + x + 1,$$

which is approximately 1.17628. Salem numbers are always algebraic units.

Let α be a Salem number. Then by definition, at least one Galois conjugate of α lies in the Julia set of $\phi(z) = z^d$. On the other hand, directly from the definition of the (canonical) height we have

$$\hat{h}_f(\alpha) = h(\alpha) = \sum_{i=1}^{d} \log \max\{|\alpha_i|, 1\} = \log \alpha,$$

since other than α itself, all of the Galois conjugates of α lie on or in the unit circle. Since α is algebraic, it is true that $\log \alpha$ is transcendental, which is good. On the other hand,

$$\exp(\hat{h}_f(\alpha)) = \exp(\log \alpha) = \alpha$$

is algebraic, despite the fact that all but two of the Galois conjugates of α are in the Julia set of ϕ.

CHAPTER 6

Postcritically Finite Maps

The orbits of the critical (or ramification) points of a rational map are important in studying the dynamics of the map. These orbits are also interesting arithmetically. For this chapter, unless we specifically say otherwise, we work over a field

$$\boxed{K \text{ of characteristic } 0.}$$

Definition 6.1. Let $\phi(z) \in \mathrm{Hom}_d$. A point $\alpha \in \mathbb{P}^1$ is a *critical point* for ϕ if

$$\phi'(\alpha) = 0.$$

This definition assumes that $\alpha \neq \infty$ and $\phi(\alpha) \neq \infty$. In general, we can make a change of coordinates to achieve these conditions. More intrinsically, the map ϕ induces a map on the one-dimensional tangent spaces

$$(\phi_\alpha)_* \colon \mathcal{T}_\alpha(\mathbb{P}^1) \longrightarrow \mathcal{T}_{\phi(\alpha)}(\mathbb{P}^1),$$

and we define

$$\phi \text{ is a critical point} \iff (\phi_\alpha)_* \text{ is the zero map;}$$

cf. Remark 2.32, which gives an intrinsic definition of the multiplier of a map at a periodic point.

Intuition. A point α is a critical point if locally around α the map ϕ looks like

$$\phi(z) = \phi(\alpha) + c(z - \alpha)^e + \cdots \quad \text{with } e \geq 2.$$

In general, if we write ϕ in this form with $c \neq 0$, then $e = e_\phi(\alpha)$ is the ramification index of ϕ at α. Thus α is a critical point if and only if $e_\phi(\alpha) \geq 2$.

Proposition 6.2. *A map $\phi \in \mathrm{Hom}_d$ has $2d - 2$ critical points, in the sense that*

$$\sum_{P \in \mathbb{P}^1} \bigl(e_\phi(P) - 1\bigr) = 2d - 2. \tag{6.1}$$

PROOF. Quotient rule. We observe that formula (6.1) is a special case of the Hurwitz genus formula, which for a finite map $C_1 \to C_2$ of curves relates the genera of C_1 and C_2 to the degree of the map and the multiplicities of the critical points. See for example [111, Theorem II.5.9]. □

Definition 6.3. A map $\phi \in \mathrm{Hom}_d$ is *postcritically finite* (PCF) if every critical point is preperiodic, i.e., if $\mathcal{O}_\phi(\alpha)$ is finite for every critical point α.

Example 6.4. Power maps z^d, Chebyshev polynomials, and Lattès maps are PCF.

6.1. Transversality of the PCF locus

Suppose that every critical point of $\phi \in \mathrm{Hom}_d$ is a fixed point. This gives $2d-2$ conditions on the coefficients of ϕ, since there are $2d-2$ critical points $c_1, \ldots, c_{2d-2}$ and we are requiring that $\phi(c_i) = c_i$ for all $1 \le i \le 2d-2$. But $\dim \mathsf{M}_d = 2d-2$, so there should be only finitely many such maps in M_d. This is indeed true, but not so easy to prove.

Definition 6.5. More generally, the *critical point portrait* of a PCF map ϕ is the directed graph whose vertices are the points in the orbits of the critical points of ϕ (assigned with multiplicity) and whose directed edges are defined by ϕ, i.e., there is an arrow $P \to P'$ if and only if $\phi(P) = P'$. Each vertex has exactly one outgoing arrow, but may have many incoming arrows.

Example 6.6. Here is a typical critical point portrait for a degree 3 map ϕ. The critical points of ϕ are the points P, Q, R, S.

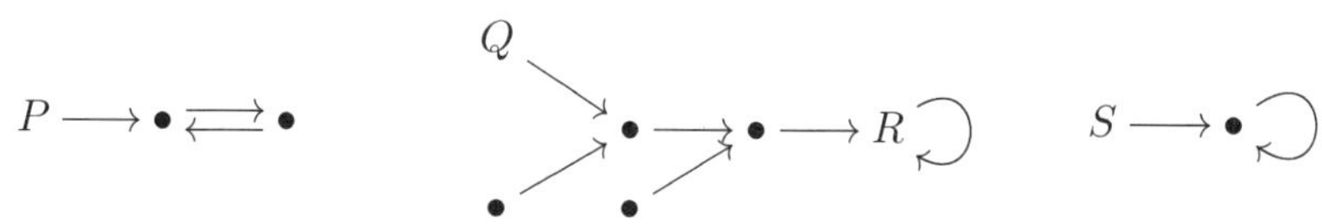

It is easy to see that a given critical point portrait leads naturally to $2d-2$ conditions on the coefficients of ϕ. However, it is not at all clear that these conditions are independent. The fact that they are independent is a consequence of Thurston's rigidity theorem.

Theorem 6.7 (Thurston [36]). *Let $d \ge 2$. There are only finitely many non-Lattès PCF maps in M_d with a given critical point portrait. In particular, the set of non-Lattès PCF maps in M_d is a countable subset of $\mathsf{M}_d(\overline{\mathbb{Q}})$.*

Even more is true. The conditions defining different critical point behaviors define subvarieties of moduli space that intersect transverally. Before stating a version of Thurston's transversality theorem, we first define a moduli space that classifies maps and their critical points listed in a specified order.

Definition 6.8. Consider the space

$$\left\{ (\phi, c_1, \ldots, c_{2d-2}) : \begin{array}{l} \phi \in \mathrm{Hom}_d \text{ and } c_1, \ldots, c_{2d-2} \text{ are the critical} \\ \text{points of } \phi \text{, listed with appropriate multiplicity} \end{array} \right\}.$$

There is a natural action of PGL_2 on this space given by

$$(\phi, c_1, \ldots, c_{2d-2})^f = \bigl(\phi^f, f^{-1}(c_1), \ldots, f^{-1}(c_{2d-2})\bigr) \quad \text{for } f \in \mathrm{PGL}_2.$$

We denote the quotient space by $\mathsf{M}_d^{\mathrm{crit}}$. It is the moduli space of maps of degree d with marked critical points.

The natural map

$$\mathsf{M}_d^{\mathrm{crit}} \longrightarrow \mathsf{M}_d, \quad (\phi, c_1, \ldots, c_{2d-2}) \longmapsto \phi,$$

is a quasi-finite covering map of degree $(2d-2)!$, since the map ϕ determines its critical points as an unordered set.

Definition 6.9. Let $\phi \in \operatorname{Hom}_d$ and let $n, r \geq 1$ be integers. We define

$$\operatorname{Per}_n^*(\phi) = \{P \in \mathbb{P}^1 : P \text{ is a point of formal period } n \text{ for } \phi\},$$
$$\operatorname{PrePer}_{r,n}^*(\phi) = \{P \in \mathbb{P}^1 : P \notin \operatorname{Per}(\phi) \text{ and } \phi^r(P) \in \operatorname{Per}_n^*(\phi)\}.$$

(See Definition 4.3 in Section 4.1 for the definition of the formal period of a point.) Thus points in $\operatorname{PrePer}_{r,n}^*(\phi)$ are preperiodic points with tail length r and formal period length n. For notational convenience, when $r = 0$ we set $\operatorname{PrePer}_{0,n}^*(\phi) = \operatorname{Per}_n^*(\phi)$.

Definition 6.10. Let $n \geq 1$ and $r \geq 0$ be integers. For each $1 \leq i \leq 2d-2$ we define a subvariety of $\mathsf{M}_d^{\text{crit}}$ by

$$\mathsf{M}_d^{\text{crit}}[i](r,n) = \{(\phi, c_1, \ldots, c_{2d-2}) \in \mathsf{M}_d^{\text{crit}} : c_i \in \operatorname{PrePer}_{r,n}^*(\phi)\}.$$

Thus $\mathsf{M}_d^{\text{crit}}[i](r,n)$ describes the set of maps of degree d with marked critical points such that the ith critical point is preperiodic with tail length r and formal period n.

Theorem 6.11 (Thurston [36]). *Fix integers $d \geq 2$, $n_1, \ldots, n_{2d-2} \geq 1$, and $r_1, \ldots, r_{2d-2} \geq 0$.*

(a) *The intersection*

$$\bigcap_{i=1}^{2d-2} \mathsf{M}_d^{\text{crit}}[i](r_i, n_i)$$

is the union of flexible Lattès maps and a finite set of points.

(b) *Let*

$$(\phi, c_1, \ldots, c_{2d-2}) \in \bigcap_{i=1}^{2d-2} \mathsf{M}_d^{\text{crit}}[i](r_i, n_i) \tag{6.2}$$

be a point with ϕ not a flexible Lattès map and with $c_1, \ldots, c_{2d-2}$ distinct. Then the subvarieties

$$\mathsf{M}_d^{\text{crit}}[1](r_1, n_1), \ \ldots, \ \mathsf{M}_d^{\text{crit}}[2d-2](r_{2d-2}, n_{2d-2})$$

intersect transversally at $(\phi, c_1, \ldots, c_{2d-2})$.

Remark 6.12. The intersection in (6.2) parametrizes the PCF rational maps having a critical point portrait in which the critical points are distinct and have orbits with specified tail length and formal period. There are, of course, other portraits in which some of the critical points appear with higher multiplicity, but we are content to state Thurston's result for the generic case described in Theorem 6.11.

Example 6.13. We illustrate Theorem 6.11 for $\phi \in \operatorname{Hom}_2$. We will assume that ϕ has a noncritical fixed point and that its two critical points are distinct. Moving the fixed point to ∞ and the critical points to 0 and -2, we find that ϕ has the form

$$\phi(z) = \frac{uz^2 + vz + v}{z+1} \quad \text{with} \quad \operatorname{Crit}(\phi) = \{0, -2\}.$$

The parameters (u, v) are affine coordinates on an open subset of $\mathsf{M}_2^{\text{crit}}$ in which the marked critical points

$$c_1 = 0 \quad \text{and} \quad c_2 = -2$$

are distinct. We can describe the subvarieties $\mathsf{M}_2^{\text{crit}}[i](r,n)$ in terms of u and v. Table 6.1 gives the equation of $\mathsf{M}_2^{\text{crit}}[1](r,n)$ and $\mathsf{M}_2^{\text{crit}}[2](r,n)$ in the cases that the

TABLE 6.1. Subvarieties of $\mathsf{M}_2^{\mathrm{crit}}$ with a (pre)periodic critical point

c_i	r	n	$\mathsf{M}_2^{\mathrm{crit}}[i](r,n)$	Portrait
0	0	1	$v=0$	
-2	0	1	$4u-v-2=0$	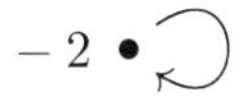
0	0	2	$uv+v+1=0$	
-2	0	2	$4u^2+2u+uv+v+1=0$	
0	1	2	$v+1=0$	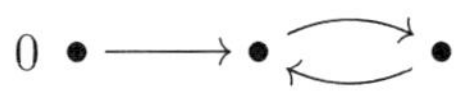
-2	1	2	$4u-v-1=0$	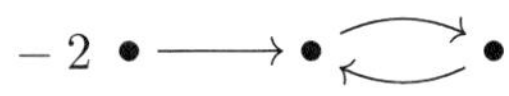

critical point c_1 or c_2 is a fixed point ($r=0$ and $n=1$), has formal period two ($r=0$ and $n=2$), or is preperiodic with tail length one and formal period two ($r=1$ and $n=2$).

For example, the equation for $\mathsf{M}_2^{\mathrm{crit}}[2](0,1)$ is obtained by setting

$$\phi(-2) = -(4u-2v+v) \quad \text{equal to } -2.$$

Similarly, the equation of $\mathsf{M}_2^{\mathrm{crit}}[2](1,2)$ is found by setting $\phi^3(-2)=\phi(-2)$, but we need to eliminate the locus where $\phi^2(-2)=-2$, so we set

$$\frac{\phi^3(-2)-\phi(-2)}{(\phi^2(-2)+2)^2} = \frac{-4u^2+uv+u}{16u^3-8vu^2+uv^2-4uv-4u+v^2+2v+1} = 0.$$

This locus also includes the set with $u=0$, which is not in our space, since $\deg(\phi)=1$ when $u=0$. So $\mathsf{M}_2^{\mathrm{crit}}[2](1,2)$ is defined by the equation $4u-v-1=0$.

The curves $\mathsf{M}_2^{\mathrm{crit}}[1](0,1)$, $\mathsf{M}_2^{\mathrm{crit}}[2](0,1)$, $\mathsf{M}_2^{\mathrm{crit}}[1](1,2)$, and $\mathsf{M}_2^{\mathrm{crit}}[2](1,2)$ are distinct lines, so their intersections satisfy Theorem 6.11. Next we consider the intersection of $\mathsf{M}_2^{\mathrm{crit}}[1](1,2)$ and $\mathsf{M}_2^{\mathrm{crit}}[2](0,2)$, that is

$$\{v+1=0\}\cap\{4u^2+2u+uv+v+1=0\}.$$

In general, two affine curves $F(u,v)=0$ and $G(u,v)=0$ intersect transversally at all of their intersection points if and only if the three equations

$$F(u,v)=0,\quad G(u,v)=0,\quad J_{F,G}(u,v)=\det\begin{vmatrix}F_u(u,v) & F_v(u,v)\\ G_u(u,v) & G_v(u,v)\end{vmatrix}=0,$$

have no simultaneous solutions. For $\mathsf{M}_2^{\mathrm{crit}}[1](1,2)\cap\mathsf{M}_2^{\mathrm{crit}}[2](0,2)$, we have

$$J(u,v)=\det\begin{vmatrix}0 & 1\\ 8u+2+v & u+1\end{vmatrix} = -(8u+2+v),$$

and one easily checks that

$$v+1 = 4u^2+2u+uv+v+1 = 8u+2+v = 0$$

has no solutions, so $\mathsf{M}_2^{\mathrm{crit}}[1](1,2)$ and $\mathsf{M}_2^{\mathrm{crit}}[2](0,2)$ intersect transversally.

Similarly, we have

$$\mathsf{M}_2^{\mathrm{crit}}[1](0,2)\cap\mathsf{M}_2^{\mathrm{crit}}[2](0,2) = \{uv+v+1=0\}\cap\{4u^2+2u+uv+v+1=0\},$$

and

$$J(u,v) = \det \begin{vmatrix} v & u+1 \\ 8u+2+v & u+1 \end{vmatrix} = -2(u+1)(4u+1).$$

Hence $\mathsf{M}_2^{\text{crit}}[1](0,2)$ and $\mathsf{M}_2^{\text{crit}}[2](0,2)$ also intersect transversally.

The proof of Thurston's general result is difficult and uses results from complex dynamics; see [36]. However, various special cases have been proven using algebriac techniques. The first of these was originally discovered by Gleason in the early 1980s (see [33, Lemma 19.1]) and has been rediscovered, simplified, and generalized by a number of people in recent years. We give Gleason's result for periodic points; see [37, Appendix] for a nontrivial generalization to preperiodic points.

Proposition 6.14. *Let $d \ge 2$ and let $\phi_c(z) = z^d + c$.*

(a) *The critical points of of ϕ_c are 0 and the fixed point ∞.*
(b) *The critical points of ϕ_c are periodic if and only if $\phi_c^n(0) = 0$ for some $n \ge 1$.*
(c) *The polynomial*
$$\phi_c^n(0) \in \mathbb{Z}[c]$$
is separable over $\mathbb{C}$, i.e., it has distinct complex roots. (In this setting, the separability of $\phi_c^n(0)$ is transversality.)

PROOF. Parts (a) and (b) are obvious. For (c), we begin by observing that a trivial induction gives

$$\phi_c^n(0) = c^{d^{n-1}} + (\text{lower order terms}) \in \mathbb{Z}[c].$$

Next we differentiate the relation

$$\phi_c^n(0) = \left(\phi_c^{n-1}(0)\right)^d + c$$

with respect to c, which shows that there is a polynomial $\psi_n(c) \in \mathbb{Z}[c]$ such that

$$\phi_c^n(0)' = d\psi_n(c) + 1.$$

Hence

$$\begin{aligned} \operatorname{Disc}\left(\phi_c^n(0)\right) &= \operatorname{Resultant}\left(\phi_c^n(0), \phi_c^n(0)'\right) \\ &= \operatorname{Resultant}\left(c^{d^{n-1}} + \cdots, \; d\psi_n(c) + 1\right). \end{aligned}$$

We can compute the resultant as the determinant of the Sylvester matrix. In our situation, the Sylvester matrix has the form

$$\begin{bmatrix} 1 & * & * & \cdots & * & & & \\ & 1 & * & * & \cdots & * & & \\ & & \ddots & & & & \ddots & \\ & & & 1 & * & * & \cdots & * \\ d & d & \cdots & d & 1 & & & \\ & d & d & \cdots & d & 1 & & \\ & & \ddots & & & & \ddots & \\ & & & d & d & \cdots & d & 1 \end{bmatrix}, \tag{6.3}$$

where we have written $*$ to indicate an arbitrary integer entry and d to indicate entries that are divisible by d. Reducing modulo d, the matrix (6.3) is upper triangular with ones on the diagonal, so taking the determinant yields

$$\operatorname{Disc}\left(\phi_c^n(0)\right) = \operatorname{Resultant}\left(\phi_c^n(0), \phi_c^n(0)'\right) \equiv 1 \pmod{d}.$$

In particular the discriminant of $\phi_c^n(0)$ is nonzero, and hence $\phi_c^n(0)$ has distinct roots. □

Gleason's proof uses reduction modulo d. Epstein [37] has given an analogous proof for polynomials of prime power p^e that uses reduction modulo p. The key step in Epstein's result is to show that for an appropriate parametrization of the moduli space of polynomials, the critical points of PCF maps are p-adically integral. More generally, he characterizes polynomials whose critical points are p-adically bounded. Silverman [113] has given an alternative proof of 3-adic integrality for cubic polynomials. We state Epstein's result and briefly sketch the proof for cubics.

We first define $\mathsf{P}_d^{\text{crit}}$ to be the subvariety of $\mathsf{M}_d^{\text{crit}}$ corresponding to polynomial maps, i.e.,

$$\mathsf{P}_d^{\text{crit}} = \bigl\{(\phi, c_1, \dots, c_{2d-2}) \in \mathsf{M}_d^{\text{crit}} : \phi(\infty) = \infty \text{ and } c_d = \cdots = c_{2d-2} = \infty\bigr\}.$$

A map ϕ in P_d has a totally ramified fixed point at ∞, so $\mathsf{P}_d^{\text{crit}}$ classifies polynomials of degree d with marked critical points. For $n \ge 1$ and $r \ge 0$, we define

$$\begin{aligned}\mathsf{P}_d^{\text{crit}}[i](r,n) &= \mathsf{P}_d^{\text{crit}} \cap \mathsf{M}_d^{\text{crit}}[i](r,n)\\ &= \bigl\{(\phi, c_1, \dots, c_{d-1}, \infty, \dots, \infty) \in \mathsf{P}_d^{\text{crit}} : c_i \in \operatorname{PrePer}_{r,n}^*(\phi)\bigr\}.\end{aligned}$$

Theorem 6.15. *Let d be a prime power, and let $n_1, \dots, n_{d-1} \ge 1$ be fixed integers. Then*

$$\mathsf{P}_d^{\text{crit}}[1](0,n_1), \dots, \mathsf{P}_d^{\text{crit}}[d-1](0,n_{d-1})$$

intersect transversally in $\mathsf{P}_3^{\text{crit}}$ at all of their points of intersection.

PROOF SKETCH FOR $d = 3$. To ease notation, we let $n = n_1$ and $m = n_2$, so we are looking at the intersection of the curves $\mathsf{P}_3^{\text{crit}}[1](0,n)$ and $\mathsf{P}_3^{\text{crit}}[2](0,m)$. Making a change of variables, we write a cubic polynomial in the form

$$\phi_{x,y}(z) = z^3 - 3x^2 z + y.$$

The critical points of $\phi_{x,y}$ are $c_1 = x$ and $c_2 = -x$ (as well as the critical fixed point at ∞). Define

$$F^{(n)}(x,y) = \phi_{x,y}^n(x) - x \quad \text{and} \quad G^{(m)}(x,y) = \phi_{x,y}^m(-x) + x.$$

The equations $F^{(n)}(x,y) = 0$ and $G^{(m)}(x,y) = 0$ define, respectively, the curves $\mathsf{P}_3^{\text{crit}}[1](0,n)$ and $\mathsf{P}_3^{\text{crit}}[2](0,m)$, so we need to check that at all intersection points of these curves, the Jacobian

$$J(x,y) = \det \begin{pmatrix} F_x^{(n)}(x,y) & G_x^{(m)}(x,y) \\ F_y^{(n)}(x,y) & G_y^{(m)}(x,y) \end{pmatrix}$$

does not vanish.

Step I. All solutions to $F^{(n)} = G^{(m)} = 0$ are 3-adically integral.
Step II. $J(x,y) \equiv 1 \pmod 3$.

Assume for the moment that we have proven Step I. Then to prove Step II, we start with the congruences

$$\begin{aligned} F^{(n)}(x,y) &\equiv x^{3^n} - x + y + y^3 + y^9 + \cdots + y^{3^{n-1}} \pmod 3,\\ G^{(m)}(x,y) &\equiv -x^{3^m} + x + y + y^3 + y^9 + \cdots + y^{3^{m-1}} \pmod 3, \end{aligned}$$

which are easily proven by induction on n and m. Step II is then an immediate consequence, since the congruences imply that

$$J(x,y) = \det \begin{pmatrix} F_x^{(n)}(x,y) & G_x^{(m)}(x,y) \\ F_y^{(n)}(x,y) & G_y^{(m)}(x,y) \end{pmatrix} \equiv \det \begin{pmatrix} -1 & 1 \\ 1 & 1 \end{pmatrix} \equiv 1 \pmod 3.$$

Hence if (α,β) is a common root of $F^{(n)}$ and $G^{(m)}$, then from Step I we know that α and β are 3-adically integral, and we also know that $J(x,y) = 1 + 3K(x,y)$ for some $K \in \mathbb{Z}[x,y]$. Therefore

$$\mathsf{N}_{\mathbb{Q}(\alpha,\beta)/\mathbb{Q}} J(\alpha,\beta) = \mathsf{N}_{\mathbb{Q}(\alpha,\beta)/\mathbb{Q}}\bigl(1 + 3K(\alpha,\beta)\bigr) \equiv 1 \pmod 3,$$

which proves that $J(\alpha,\beta) \ne 0$.

The proof of Step I is considerably more complicated. Epstein [37] gives a detailed and delicate analysis of the sequence of 3-adic valuations $\operatorname{ord}_3\bigl(\phi^n(c)\bigr)$ of the points in the forward orbit of a critical point. This proof uses the 3-adic metric and has a p-adic dynamical flavor. A second proof, due to Silverman [113], is more algebraic in nature. First one analyzes the iterates $\phi_{x,y}^n(x)$, considered as polynomials in x with coefficients in $\mathbb{Z}[y]$. More precisely, writing

$$f_{x,y}^n(x) = \sum_{k=0}^{3^n} a_k(y) x^{3^n-k} \in \mathbb{Z}[y][x],$$

one shows that

$$\deg a_k(y) \le 4\left\lfloor \frac{k}{3} \right\rfloor - k, \quad a_0(y) \equiv 1 \pmod 3, \quad \text{and} \quad a_{3^n}(y) = y^{3^{n-1}} + (\text{l.o.t.}).$$

These formulas and an expansion of the Sylvester matrix for the resultant are then used to prove that

$$\operatorname{Resultant}_x\bigl(F^{(n)}(x,y), G^{(m)}(x,y)\bigr) \in \mathbb{Z}[y] \tag{6.4}$$

is a polynomial of degree 3^{n+m-1} with integer coefficients and leading coefficient relatively prime to 3. Finally, if (α,β) is a common solution to $F^{(n)} = G^{(m)} = 0$, then β is a root of the polynomial (6.4), hence β is 3-integral, and then the fact that α is a root of $F^{(n)}(x,\beta)$ shows that α is also 3-integral. □

Remark 6.16. There is a trick to show transversality for cubic polynomials if one or both of the critical points has tail length exactly one. The proof uses the elementary identity

$$\phi_{x,y}(z) - \phi_{x,y}(x) = (z-x)^2(z+2x) \equiv (z-x)^3 \pmod 3$$

with $z = \phi_{x,y}^n(x)$. The method does not work for longer tail lengths, nor for polynomials of higher degree. See [113, §3] for details.

Remark 6.17. Thurston's tranversality property is true over $\mathbb{C}$, so it is true over any field of characteristic 0 by the Lefschetz principle. However, it is not necessarily true over fields of characteristic p. We consider the simplest case of the quadratic polynomial $\phi_c(z) = z^2 + c$. As noted in the statement of Proposition 6.14, transversality in this case is the statement that the roots of $\phi_c^n(0)$ are distinct. So transversality fails for primes dividing the discriminant of $\phi_c^n(0)$. We list in Table 6.2 the first few values, where p_r indicates an r-digit prime and q_r indicates an r-digit number with no prime factors smaller than 10^9. It would be interesting to understand the primes for which transversality fails. Here are two specific questions:

TABLE 6.2

n	$\mathrm{Disc}\big(\phi_c^n(0)\big)$
1	1
2	1
3	-23
4	$23 \cdot 2551$
5	$13 \cdot p_{17}$
6	$13 \cdot 23 \cdot 949818439 \cdot p_{28}$
7	$-8291 \cdot 9137 \cdot 420221 \cdot q_{99}$

- Fix a prime p. Consider the set

$$\big\{n : \mathrm{Disc}\big(\phi_c^n(0)\big) \equiv 0 \pmod{p}\big\}. \tag{6.5}$$

It is clear that if n is in the set (6.5), then so is nk for all $k \geq 1$. (This follows from the fact that $\phi_c^{nk}(0)$ divides $\phi_c^n(0)$ in $\mathbb{Z}[c]$.) Is there a finite set of integers $\{n_1, \ldots, n_r\}$ such that the set (6.5) is the union of $n_i\mathbb{N}$?

- How large is the set

$$\big\{p : \text{there is an } n \geq 1 \text{ such that } \mathrm{Disc}\big(\phi_c^n(0)\big) \equiv 0 \pmod{p}\big\}? \tag{6.6}$$

In particular, does it have density zero? We mention that in number theoretic terminology, the set (6.6) is called the *support* of the sequence $\big(\mathrm{Disc}\big(\phi_c^n(0)\big)\big)_{n\geq 0}$.

One would expect that the PCF locus should be Zariski dense in M_d, but this fact does not seem to appear in the literature. We sketch a proof shown to us by Adam Epstein.

Proposition 6.18 (Epstein, private communication). *The postcritically finite maps are Zariski dense in the moduli space* M_d.

PROOF SKETCH. Let $\phi \in \mathrm{Hom}_d$ be a non-Lattès PCF map whose critical points are distinct and strictly preperiodic. Buff and Epstein [23] have shown that in a neighborhood of ϕ, the loci in M_d where the various relations on the critical orbit persist are pairwise transverse smooth hypersurfaces. Suitable intersections of these spaces yield curves whose tangent spaces at ϕ collectively span the tangent space to M_d at ϕ. Then a standard (Montel's theorem) argument of Mañé–Sad–Sullivan shows that on each such curve, the map ϕ is a limit point of PCF maps lying on the curve. Hence any subvariety containing the PCF locus has dimension equal to the dimension of M_d, so is equal to M_d, since M_d is irreducible. □

Given a projective variety X and a Zariski dense list of points $(P_1, P_2, \ldots)$ of X, one can ask to what extent the points are in general position. The following definition is used to make this notion precise.

Definition 6.19. Let $X \subset \mathbb{P}^N$ be a projective variety. For any finite set of points $S \subset X$, define $\deg_X(S)$ to be the minimal degree of a homogeneous polynomial $F \in k[x_0, \ldots, x_N]$ that does not vanish identically on X, but vanishes at all of the points in S. Equivalently, $\deg_X(S)$ is the smallest integer $d \geq 1$ with the property that

$$H^0\big(\mathbb{P}^N, \mathcal{I}_S(d)\big) \smallsetminus H^0\big(\mathbb{P}^N, \mathcal{I}_X(d)\big) \neq \varnothing, \tag{6.7}$$

where in general $\mathcal{I}_V$ is the ideal sheaf associated to the variety $V \subset \mathbb{P}^N$. (Question: Is $\deg_X(S)$ independent of the projective embedding of X?)

If d is sufficiently large, so that $H^1(\mathbb{P}^N, \mathcal{I}_X(d)) = 0$, then an easy calculation shows that

$$h^0(\mathbb{P}^N, \mathcal{I}_S(d)) - h^0(\mathbb{P}^N, \mathcal{I}_X(d)) \geq h^0(X, \mathcal{O}_X(d)) - \# S.$$

Further, the Hilbert polynomial of X has degree equal to the dimension of X, so

$$h^0(X, \mathcal{O}_X(d)) \sim cd^{\dim X}$$

for sufficiently large d and some $c > 0$. Hence if

$$(\dim X) \cdot (\log d) > \log \# S,$$

then (6.7) is true, which shows that

$$(\dim X) \cdot (\log \deg_X(S)) \leq \log \# S$$

as $\# S \to \infty$. This prompts the following definition.

Definition 6.20. Let $X \subset \mathbb{P}^N$ be a projective variety, let $S_1 \subset S_2 \subset S_3 \subset \cdots$ be a sequence of finite subsets of X, and let $S = \bigcup S_n$. Assume that $\# S = \infty$. We will say that S is *Zariski equidistributed in* X (relative to the given filtration) if

$$\lim_{n\to\infty} \frac{\log \deg_X S_n}{\log \# S_n} = \frac{1}{\dim X}.$$

Conjecture 6.21. *Fix a projective embedding $\mathsf{M}_d \subset \mathbb{P}^N$. For each integer $n \geq 1$, let $S_n^{\mathrm{crit}} \subset \mathsf{M}_d$ be the set of non-Lattès maps ϕ of degree d such that there are at most n points in the union of the critical orbits of ϕ. (In particular, ϕ is postcritically finite.) Then $S^{\mathrm{crit}} = \bigcup S_n^{\mathrm{crit}}$ is Zariski equidistributed in M_d.*

Definition 6.22. A *partial PCF portrait* is a portrait that specifies an orbit configuration for a subset of the critical points. For example, a partial PCF portrait is associated to the intersection of one or moreof sets of the form $\mathsf{M}_d^{\mathrm{crit}}[i](r_i, n_i)$ as described in Definition 6.10.

Question 6.23. We close this section with two questions.

(a) For a given partial PCF portrait, is the associated subvariety of $\mathsf{M}_d^{\mathrm{crit}}$ irreducible?
(b) For a given partial PCF portrait, is the associated subvariety of $\mathsf{M}_d^{\mathrm{crit}}$ nonsingular?

Epstein and Buff (private communication) have noted that over $\mathbb{C}$, the answer to (b) is yes and follows from Thurston transversality (Theorem 6.11), but that the answer to (a) is not known. They also observe that since (b) is true, irreducibility is equivalent to connectedness, but that it is not known if the partial portrait subvarieties of dimension at least one are connected. The problem is that partial portrait subvarieties live in the affine variety $\mathsf{M}_d^{\mathrm{crit}}$, so there could be points missing at infinity that disconnect them. It would also be interesting to investigate these questions in characteristic p.

6.2. The height of a postcritically finite map

It is well-known that the behavior of the critical points characterizes much of the dynamics of a map. This suggests the following arithmetic definition.

Definition 6.24. The *critical canonical height* of a map $\phi \in \operatorname{Hom}_d(\overline{\mathbb{Q}})$ is the quantity

$$\hat{h}_{\mathrm{crit}}(\phi) = \sum_{P \in \operatorname{Crit}(\phi)} \hat{h}_\phi(P).$$

Of course, each critical point must be taken in the sum with appropriate multiplicity, so an equivalent definition is

$$\hat{h}_{\mathrm{crit}}(\phi) = \sum_{P \in \mathbb{P}^1} \bigl(e_\phi(P) - 1\bigr)\hat{h}_\phi(P).$$

The function $\hat{h}_{\mathrm{crit}}$ is PGL_2 invariant, so

$$\hat{h}_{\mathrm{crit}} : \mathsf{M}_d(\overline{\mathbb{Q}}) \longrightarrow \mathbb{R}$$

is a function that measures the arithmetic dynamical complexity of the critical points of a map.

We start with two basic properties of the critical height.

Proposition 6.25. *Let $\phi \in \operatorname{Hom}_d(\overline{\mathbb{Q}})$.*

(a) *The map ϕ is PCF if and only if $\hat{h}_{\mathrm{crit}}(\phi) = 0$.*
(b) *The critical height of an iterate is given by the formula*

$$\hat{h}_{\mathrm{crit}}(\phi^n) = n\hat{h}_{\mathrm{crit}}(\phi).$$

PROOF. (a) It is clear from the definition that $\hat{h}_\phi(P)$ is nonnegative, so $\hat{h}_{\mathrm{crit}}(\phi) = 0$ if and only if $\hat{h}_\phi(P) = 0$ for all $P \in \operatorname{Crit}(\phi)$. Now we apply Proposition 5.6, which says that $\hat{h}_\phi(P) = 0$ if and only if P is preperiodic.

(b) The chain rule implies that

$$\operatorname{Crit}(\phi^n) = \bigcup_{i=0}^{n-1} \phi^{-i}\bigl(\operatorname{Crit}(\phi)\bigr), \tag{6.8}$$

so

$$\begin{aligned}
\hat{h}_{\mathrm{crit}}(\phi^n) &= \sum_{P \in \operatorname{Crit}(\phi^n)} \hat{h}_{\phi^n}(P) && \text{definition of } \hat{h}_{\mathrm{crit}},\\
&= \sum_{P \in \operatorname{Crit}(\phi^n)} \hat{h}_{\phi}(P) && \text{since } \hat{h}_{\phi^n} = \hat{h}_\phi,\\
&= \sum_{i=0}^{n-1} \sum_{P \in \operatorname{Crit}(\phi)} \sum_{Q \in \phi^{-i}(P)} \hat{h}_\phi(Q) && \text{from (6.8)},\\
&= \sum_{i=0}^{n-1} \sum_{P \in \operatorname{Crit}(\phi)} \sum_{Q \in \phi^{-i}(P)} d^{-i}\hat{h}_\phi(P) && \text{since } \hat{h}_\phi \circ \phi^i = d^i\hat{h}_\phi,\\
&= \sum_{i=0}^{n-1} \sum_{P \in \operatorname{Crit}(\phi)} \hat{h}_\phi(P) && \text{since } \#\,\phi^{-i}(P) = d^i,\\
&= n\hat{h}_{\mathrm{crit}}(\phi) && \text{definition of } \hat{h}_{\mathrm{crit}}.
\end{aligned}$$

□

It is always of interest to ask how small a canonically defined height can be if it is nonzero. The first such question was raised by Lehmer, who asked if there is an absolute constant $C > 0$ such that if $\alpha \in \overline{\mathbb{Q}}^*$ is not a root of unity, then $h(\alpha) > C/[\mathbb{Q}(\alpha) : \mathbb{Q}]$. Faber, Ghioca, and Tucker have posed the following PCF analogue.

Question 6.26. Is there a constant $C(d) > 0$ such that

$$\hat{h}_{\text{crit}}(\phi) > C(d) \quad \text{for all } \phi \in M_d(\overline{\mathbb{Q}}) \text{ satisfying } \hat{h}_{\text{crit}}(\phi) \neq 0?$$

Might this even be true for an absolute constant C that is independent of d?

Remark 6.27. For maps $\phi \colon \mathbb{P}^1(\mathbb{C}) \to \mathbb{P}^1(\mathbb{C})$, we can form a sum using Green functions that is analogous to the sum defining $\hat{h}_{\text{crit}}$. The function

$$\hat{g}_{\text{crit}}(\phi) = \sum_{P \in \text{Crit}(\phi)} \hat{g}_\phi(P)$$

has been extensively studied in complex dynamics and is closely related to the Lyapunov exponent of the map. Here $\hat{g}_\phi$ is the complex Green function (Section 5.4) associated to the Julia set of ϕ. See for example [30, 92]; the latter article also considers the non-Archimedean case.

Definition 6.28. We recall the standard notation

$$G(x) \ll F(x),$$

which means that there are positive constants c_1, c_2 such that

$$G(x) \leq c_1 F(x) + c_2.$$

We also write $F(x) \asymp G(x)$ if both $G(x) \ll F(x)$ and $F(x) \ll G(x)$.

Conjecture 6.29. *Fix a height function h_{M_d} on M_d corresponding to some projective embedding, and let $\text{Lat}_d \subset \text{Hom}_d$ denote the locus of flexible Lattès maps. Then*

$$h_{\mathsf{M}_d}(\phi) \asymp \hat{h}_{\text{crit}}(\phi) \qquad \textit{for } \phi \in (\text{Hom}_d \smallsetminus \text{Lat}_d)(\overline{\mathbb{Q}}),$$

where the implied constants depend on d and on the choice of the height function h_{M_d}.

As noted earlier in Remark 3.21, if d is not a perfect square, then the set of flexible Lattès maps Lat_d is empty. An immediate consequence of Conjecture 6.29 would be a bound for the height of postcritically finite maps.

Conjecture 6.30. *Let L_d denote the image of Lat_d in M_d. Then*

$$\bigl\{\phi \in (\mathsf{M}_d \smallsetminus \mathsf{L}_d)(\overline{\mathbb{Q}}) : \phi \textit{ is PCF}\bigr\} \quad \textit{is a set of bounded height.}$$

One direction of Conjecture 6.29 is known. Notice that in this direction, there is no need to exclude the Lattès maps.

Theorem 6.31. *With notation as in Conjecture* 6.29,

$$h_{\mathsf{M}_d}(\phi) \gg \hat{h}_{\text{crit}}(\phi) \quad \textit{for } \phi \in \text{Hom}_d(\overline{\mathbb{Q}}). \tag{6.9}$$

PROOF. To ease notation, we write $h_{\text{Hom}}(\phi)$ for the Weil height of the point $\phi \in \text{Hom}_d \subset \mathbb{P}^{2d+1}$, and we write h_{M} for the height associated to a fixed projective embedding of $\mathsf{M}_d \subset \mathbb{P}^N$. We note that $\hat{h}_{\text{crit}}(\phi)$ is invariant under the natural action of PGL_2, so $\hat{h}_{\text{crit}}$ gives a well-defined function on $\mathsf{M}_d(\overline{\mathbb{Q}})$. The same is obviously true of h_{M}, so in proving the desired estimate (6.9), we are free to replace ϕ by any map that is PGL_2-conjugate to ϕ. Before continuing with the proof, we quote a general lemma relating heights and flat morphisms, which we will then apply to the map $\mathrm{Hom}_d \to \mathsf{M}_d$.

Lemma 6.32. *Let X and Y be quasi-projective varieties defined over a global field K, let $\mathcal{L}$ and $\mathcal{M}$ be ample line bundles on X and Y, respectively, and let*

$$F\colon X \longrightarrow Y$$

be a flat surjective morphism. Define a function

$$h_{F,\mathcal{L}}^{\min}\colon Y(\overline{\mathbb{Q}}) \longrightarrow \mathbb{R}, \quad h_{F,\mathcal{L}}^{\min}(y) = \min_{x\in F^{-1}(y)} h_{X,\mathcal{L}}(x).$$

Then

$$h_{F,\mathcal{L}}^{\min} \asymp h_{Y,\mathcal{M}}$$

PROOF. The proof is a double induction on the dimension $\dim(X)$ and the relative dimension $\dim(X) - \dim(Y)$. For lack of a suitable reference, we give a detailed sketch of the proof. The estimate

$$h_{Y,\mathcal{M}}\bigl(F(x)\bigr) \ll h_{X,\mathcal{L}}(x) \quad \text{for all } x \in X(\overline{K}) \tag{6.10}$$

follows immediately from the triangle inequality, so in particular

$$\begin{aligned} h_{F,\mathcal{L}}^{\min}(y) &= \min_{x\in F^{-1}(y)} h_{X,\mathcal{L}}(x) && \text{by definition,}\\ &\gg \min_{x\in F^{-1}(y)} h_{Y,\mathcal{M}}(F(x)) && \text{from (6.10),}\\ &= h_{Y,\mathcal{M}}(y). \end{aligned}$$

This gives one direction.

For the other direction, we induct on the relative dimension

$$\dim(X/Y) = \dim(X) - \dim(Y).$$

Consider first the case $\dim(X/Y) = 0$. The flatness of F then implies that F is quasi-finite, i.e., it has finite fibers. In this relative dimension zero case, we will prove the stronger result

$$h_{X,\mathcal{L}} \asymp h_{Y,\mathcal{M}} \circ F. \tag{6.11}$$

The proof is by a second induction on $\dim(X) = \dim(Y)$. There's nothing to prove for $\dim(X) = 0$, since X and Y are finite sets of points. Assume now that we have proven the result for $\dim(X) < n$, and let $\dim(X) = n$. The morphism $F\colon X \to Y$ is, in particular, a dominant equidimensional rational map, so the main result of [112] gives the desired result off of a closed subset, i.e., there is a proper Zariski closed subset $W \subset X$ such that

$$h_{X,\mathcal{L}}(x) \asymp h_{Y,\mathcal{M}}\bigl(F(x)\bigr) \quad \text{for all } x \in (X \smallsetminus W)(\overline{K}). \tag{6.12}$$

We now consider the flat morphism

$$F|_W\colon W \longrightarrow F(W).$$

The induction hypothesis applied to $F|_W$ give

$$h_{W,\mathcal{L}|_W}(x) \asymp h_{F(W),\mathcal{M}|_{F(W)}}\bigl(F(x)\bigr) \quad \text{for all } x \in W(\overline{K}). \tag{6.13}$$

Combining (6.12) and (6.13) gives the desired result (6.11) in the case that $\dim(X/Y) = 0$.

Assume now that we have proven the proposition for $\dim(X/Y) < n$ for some $n \geq 1$, and we take varieties with $\dim(X/Y) = n$. For a large value of d, consider the subvarieties of X defined by global sections of $\mathcal{O}_X(d)$. For most $f \in H^0(X, \mathcal{O}_X(d))$, the restriction of the map F to the subvariety variety $V = V(f) = \{x \in X : f(x) = 0\}$ is flat and surjective. We denote this map by

$$F_V : V \longrightarrow Y.$$

(This is where we are using the assumption that $\dim(X/Y) \geq 1$, otherwise F_V would never be surjective, since $\dim(V)$ would be smaller than $\dim(Y)$.) Applying the induction hypothesis to V, we have

$$\begin{aligned}
h_{F,\mathcal{L}}(y) &= \min_{x \in F^{-1}(y)} h_{X,\mathcal{L}}(x) \\
&\leq \min_{x \in F^{-1}(y) \cap V} h_{X,\mathcal{L}}(x) && \text{taking min over smaller set,} \\
&= \min_{x \in F_V^{-1}(y)} h_{V,\mathcal{L}|_V}(F(x)) && \text{since } V \subset X, \\
&= h_{F_V,\mathcal{L}|_V}(y) && \text{definition of } h_{F,\mathcal{L}}, \\
&\ll h_{Y,\mathcal{M}}(y) && \text{by the induction hypothesis.}
\end{aligned}$$

This completes the proof of the lemma. □

We apply Lemma 6.32 to the flat surjective map

$$\mathrm{Hom}_d \longrightarrow \mathsf{M}_d.$$

This allows us to replace ϕ with a conjugate map (which by abuse of notation we also denote by ϕ) satisfying

$$h_{\mathrm{Hom}}(\phi) \asymp h_{\mathsf{M}}(\phi), \tag{6.14}$$

where the implied constants depend only on d.

As noted in the statement of Theorem 5.5, the dependence of the $O(1)$ on the map ϕ in the estimate for the difference between the Weil height and the canonical height can be given explicitly as

$$\hat{h}_\phi(P) = h(P) + O_d\bigl(h_{\mathrm{Hom}}(\phi) + 1\bigr) \quad \text{for all } P \in \mathbb{P}^1(\overline{\mathbb{Q}}) \text{ and } \phi \in \mathrm{Hom}_d(\overline{\mathbb{Q}}). \tag{6.15}$$

Now consider the parameter space of rational maps with marked critical points,

$$\mathrm{Hom}_d^{\mathrm{crit}} = \left\{ (\phi; P_1, \ldots, P_{2d-2}) : \begin{matrix} \phi \in \mathrm{Hom}_d \text{ and} \\ \mathrm{Crit}(\phi) = \{P_1, \ldots, P_{2d-2}\} \end{matrix} \right\}.$$

(Here $\{P_1, \ldots, P_{2d-2}\}$ is a multiset, i.e., a set in which elements may appear multiple times.) We put a height on $\mathrm{Hom}_d^{\mathrm{crit}}$ via the natural embedding

$$\mathrm{Hom}_d^{\mathrm{crit}} \subset \mathrm{Hom}_d \times \mathbb{A}^{2d-2} \subset \mathbb{P}^{2d+1} \times (\mathbb{P}^1)^{2d-2}.$$

The projection

$$\mathrm{Hom}_d^{\mathrm{crit}} \longrightarrow \mathrm{Hom}_d, \quad (\phi; P_1, \ldots, P_{2d-2}) \longmapsto \phi,$$

is quasi-finite, so standard height estimates give

$$h(\phi; P_1, \ldots, P_{2d-2}) \asymp h(\phi).$$

Since

$$h(\phi; P_1, \ldots, P_{2d-2}) = h(\phi) + \sum_{i=1}^{2d-2} h(P_i),$$

we obtain

$$\sum_{P\in\mathrm{Crit}(\phi)} h(P) \ll h(\phi). \tag{6.16}$$

Combining the above estimates allows us to compute

$$\begin{aligned}
\hat{h}_{\mathrm{crit}}(\phi) &= \sum_{P\in\mathrm{Crit}(\phi)} \hat{h}_\phi(P) && \text{definition of } \hat{h}_{\mathrm{crit}},\\
&= \sum_{P\in\mathrm{Crit}(\phi)} h(P) + O_d\bigl(h(\phi)+1\bigr) && \text{using (6.15)},\\
&\ll h(\phi) && \text{from (6.16)},\\
&\ll h_{\mathsf{M}}(\phi) && \text{from (6.14)}.
\end{aligned}$$

This concludes the proof of Theorem 6.31. □

Patrick Ingram has proven the other direction of Conjecture 6.29 for polynomials.

Theorem 6.33 (Ingram [53]). *Conjecture* 6.29 *is true for polynomials, i.e.,*

$$h_{\mathsf{M}}(\phi) \asymp \hat{h}_{\mathrm{crit}}(\phi) \quad \textit{for } \phi \in \overline{\mathbb{Q}}[z].$$

PROOF SKETCH. Every polynomial ϕ of degree d is conjugate via an affine transformation $z \to \alpha z + \beta$ to a normalized polynomial of the form

$$z^d + a_{d-2}z^{d-2} + \cdots + a_1 z + a_0,$$

and ϕ determines $(a_{d-2},\ldots,a_0)$ up to a transformation $a_i \to \zeta^{i-1}a_i$ for some $\zeta \in \boldsymbol{\mu}_{d-1}$. Then

$$h_{\mathsf{M}_d}(\phi) = h\bigl([1, a_{d-2},\ldots,a_0]\bigr)$$

gives a height function on the set of polynomial maps in M_d.

For any $\mathbf{c} = (c_1,\ldots,c_{d-1}) \in \mathbb{A}^{d-1}(\overline{\mathbb{Q}})$ we let

$$\phi_{\mathbf{c}}(z) = \int_0^z \prod_{i=1}^{d-1}(z-c_i)\,dz = \sum_{i=0}^{d-1} \frac{(-1)^i s_i}{d-i} z^{d-i},$$

where $s_0,\ldots,s_{d-1}$ are the elementary symmetric polynomials in $c_1,\ldots,c_{d-1}$. Notice that

$$\phi'_{\mathbf{c}}(z) = \prod_{i=1}^{d-1}(z-c_i),$$

so $c_1,\ldots,c_{d-1}$ are the critical points of $\phi_{\mathbf{c}}(z)$.

A standard estimate relating the height of a polynomial to the height of its roots [111, VIII.5.9], together with the elementary estimate $h(F) = h(F') + O(1)$ for polynomials satisfying $F(0) = 0$, gives

$$h_{\mathsf{M}_d}(\phi_{\mathbf{c}}) \asymp h\bigl([1, c_1,\ldots,c_{d-1}]\bigr). \tag{6.17}$$

We now need to relate the naive heights of the critical points to their $\phi_{\mathbf{c}}$-canonical heights. In general there is an estimate $h(P) = \hat{h}_\phi(P) + O(1)$, but this is not immediately useful because the $O(1)$ depends on ϕ. More precisely, $h(P) = \hat{h}_\phi(P) + O(h(\phi)+1)$ with constants depending on $\deg(\phi)$.

For any i, the quantity $\phi_{\mathbf{c}}(c_i)$ is a homogeneous polynomial of degree d in $\mathbb{Q}[c_1,\ldots,c_{d-1}]$. The following lemma is crucial in proving a key estimate.

Lemma 6.34. *The homogeneous polynomials* $\phi_{\mathbf{c}}(c_1), \ldots, \phi_{\mathbf{c}}(c_{d-1})$ *have no nontrivial common zeros.*

PROOF. Suppose that $(c_1, \ldots, c_{d-1})$ is a common zero, or equivalently, suppose that for all i, the polynomial $\phi_{\mathbf{c}}(z)$ vanishes at c_i. Since $\phi'_{\mathbf{c}}(z) = \prod(z - c_i)$, we see that

$$\operatorname{ord}_{z=c_i} \phi_{\mathbf{c}}(z) = 1 + \operatorname{ord}_{z=c_i} \phi'_{\mathbf{c}}(z) = 1 + \#\{j : c_j = c_i\}.$$

Relabeling the c_i, we suppose that $c_1, \ldots, c_t$ are the distinct values that appear in the list $c_1, \ldots, c_{d-1}$. Then

$$\begin{aligned} d = \deg \phi_{\mathbf{c}} \geq \sum_{i=1}^{t} \operatorname{ord}_{z=c_i} \phi_{\mathbf{c}}(z) &= \sum_{i=1}^{t} \bigl(1 + \#\{j : c_j = c_i\}\bigr) \\ &= t + \sum_{i=1}^{t} \#\{j : c_j = c_i\} \\ &= t + (d-1). \end{aligned}$$

Hence $t = 1$, so $c_1, \ldots, c_{d-1}$ all have the same value. Since $\phi_{\mathbf{c}}(0) = 0$, at least one of c_i is equal to zero, and therefore they are all zero. □

We recall from Section 5.4 that there are v-adic local canonical height functions $\hat{g}_{\phi,v}$ that sum to $\hat{h}_\phi$; see Definition 5.18. The local height satisfies $\hat{g}_{\phi,v}(\alpha) = \log^+ |\alpha|_v + O(1)$, but the $O(1)$ depends on the map ϕ. This follows from the fact that if $|\alpha|_v$ is sufficiently large in comparison to the v-adic size of the coefficients of ϕ, then $\bigl|\phi(\alpha)\bigr|_v$ will be approximately equal to $|\alpha|_v^d$, and in fact they will be exactly equal for all but finitely many v. Quantifying this observation, an elementary triangle inequality estimate gives the following result, which we state using an elegant definition due to Lang.

Definition 6.35. An $M_{\mathbb{Q}}$-constant is a map $\xi \colon M_{\overline{\mathbb{Q}}} \to \mathbb{R}$ that is consistent on extension fields and such that for any particular number field K, the set $\{v \in M_K : \xi(v) \neq 0\}$ is finite. An $M_{\mathbb{Q}}$-bounded function on a set S is a map $\epsilon \colon S \times M_{\overline{\mathbb{Q}}} \to \mathbb{R}$ such that there is an $M_{\overline{\mathbb{Q}}}$-constant satisfying $\bigl|\epsilon(\alpha, v)\bigr| \leq \xi(v)$ for all $(a, v) \in S \times M_{\overline{\mathbb{Q}}}$. (See [62] for further details.)

Lemma 6.36. *There is an* $M_{\mathbb{Q}}$*-constant* ξ_1 *and an* $M_{\mathbb{Q}}$*-bounded function* ϵ_1*, depending only on* d*, such that for all* $c_1, \ldots, c_{d-1} \in \overline{\mathbb{Q}}$*, all* $v \in M_{\overline{\mathbb{Q}}}$*, and all* $\alpha \in \mathbb{A}^1(\overline{\mathbb{Q}})$*,*

$$|\alpha|_v > \log^+ \max_{1 \leq i \leq d-1} |c_i|_v + \xi_1(v) \implies \hat{g}_{\phi_{\mathbf{c}},v}(\alpha) = \log |\alpha|_v + \epsilon_1(\alpha, v).$$

PROOF. This lemma is an amalgamation of Lemmas 2.1 and 2.2 in [53]. □

The local height satisfies $\hat{g}_{\phi,v}(\alpha) = \log^+ |\alpha|_v + O(1)$, but there may be some α for which $\bigl|\phi^n(\alpha)\bigr|_v$ stays small until n gets quite large. We use Lemmas 6.34 and 6.36 to show that for a critical point c of a normalized polynomial $\phi_{\mathbf{c}}$, the quantity $\bigl|\phi(\alpha)\bigr|_v$ is already reasonably large.

Lemma 6.37. *There is an* $M_{\mathbb{Q}}$*-constant* ξ_3 *such that for all*

$$c_1, \ldots, c_{d-1} \in \overline{\mathbb{Q}} \text{ and all } v \in M_{\overline{\mathbb{Q}}},$$

we have

$$\sum_{i=1}^{d-1} \hat{g}_{\phi,v}(c_i) \ge \frac{1}{2}\log^+ \max_{1\le i\le d-1} |c_i|_v - \xi_3(v). \tag{6.18}$$

PROOF. The fact that the homogeneous polynomials

$$\phi_{\mathbf{c}}(c_1),\ldots,\phi_{\mathbf{c}}(c_{d-1}) \in \mathbb{Z}[c_1,\ldots,c_{d-1}]$$

have no nontrivial common zeros (Lemma 6.34) can be combined with the Nullstellensatz and triangle inequality estimates to show that

$$\log \max_{1\le i\le d-1} \bigl|\phi_{\mathbf{c}}(c_i)\bigr|_v = d\log \max_{1\le i\le d-1} |c_i|_v + \epsilon_2(\mathbf{c},v), \tag{6.19}$$

where ϵ_2 is an $M_{\mathbb{Q}}$-bounded function depending only on d. (See, for example, the proof of [111, VIII.5.6].) We let $\xi_2(v) = \sup_{\mathbf{c}}\bigl|\epsilon_2(\mathbf{c},v)\bigr|$ be an $M_{\mathbb{Q}}$-constant that dominates the $M_{\mathbb{Q}}$-bounded function ϵ_2. We consider two cases. Suppose first that $\frac{1}{2}\log\max|c_i|_v \le \xi_1(v)+\xi_2(v)$. Then we have a trivial estimate (using the fact that the local heights are nonnegative)

$$\sum_{i=1}^{d-1} \hat{g}_{\phi,v}(c_i) \ge 0 \ge \frac{1}{2}\log \max_{1\le i\le d-1} |c_i|_v - \xi_1(v) - \xi_2(v).$$

So in this case we can take $\xi_3 = \xi_1 + \xi_2$.

Next we suppose that

$$\frac{1}{2}\log\max|c_i|_v > \xi_1(v)+\xi_2(v).$$

Then using the fact that $d\ge 2$ and $|\epsilon_2| \le \xi_2$, we can rewrite (6.19) as

$$\begin{aligned} \log \max_{1\le i\le d-1} \bigl|\phi_{\mathbf{c}}(c_i)\bigr| &= \left(d-\frac{1}{2}\right)\log \max_{1\le i\le d-1} |c_i|_v + \left(\frac{1}{2}\log \max_{1\le i\le d-1} |c_i|_v + \epsilon_2(\mathbf{c},v)\right) \\ &\ge \frac{3}{2}\log \max_{1\le i\le d-1} |c_i|_v + \xi_1(v). \end{aligned}$$

Taking j to be an index that maximizes $\bigl|\phi_{\mathbf{c}}(c_j)\bigr|$, this inequality allows us to apply Lemma 6.36 with $\alpha = \phi_{\mathbf{c}}(c_j)$ to get

$$\begin{aligned} d\hat{g}_{\phi_{\mathbf{c}},v}(c_j) &= \hat{g}_{\phi_{\mathbf{c}},v}\bigl(\phi_{\mathbf{c}}(c_j)\bigr) && \text{Proposition 5.19(a)},\\ &= \log\bigl|\phi_{\mathbf{c}}(c_j)\bigr|_v + \epsilon_1(\mathbf{c},v) && \text{Lemma 6.36 with } \alpha = \phi_{\mathbf{c}}(c_j),\\ &= \log \max_{1\le i\le d-1} \bigl|\phi_{\mathbf{c}}(c_i)\bigr|_v + \epsilon_1(\mathbf{c},v) && \text{from choice of } j,\\ &= d\log \max_{1\le i\le d-1} |c_i|_v + \epsilon_2(\mathbf{c},v) && \text{from (6.19).} \end{aligned}$$

So in this case we can take $\epsilon_3 = d^{-1}\epsilon_2$, which completes the proof of Lemma 6.37. □

We resume the proof of Theorem 6.33. Let $c_1,\ldots,c_{d-1} \in K$ for some number field K. Summing an appropriate multiple of (6.18) in Lemma 6.37 and using the local decomposition of the canonical height as described in Proposition 5.19(e) gives

$$\sum_{i=1}^{d-1} \hat{h}_\phi(c_i) \ge \frac{1}{2}h\bigl([1,c_1,\ldots,c_{d-1}]\bigr),$$

and finally (6.17) tells us that this last quantity is $\asymp h_{\mathsf{M}_d}(\phi_{\mathbf{c}})$. This completes the proof of Theorem 6.33. $\square$

We conclude this section with some explicit estimates for a particular one-dimensional family.

Theorem 6.38 (Rivera-Letelier, Silverman). *Fix integers $d > e \geq 1$ and consider*

$$\mathrm{PCF}(d,e) = \left\{c \in \overline{\mathbb{Q}} : \phi(z) = z^d + cz^e \text{ is PCF}\right\}.$$

(a) *Let $c \in \mathrm{PCF}(d,e)$. Then $(d-e)^{d-e}(ec/d)^{d-1}$ is an algebraic integer. In particular, c is p-integral for all primes $p \nmid (d-e)e$.*

(b) *For any μ satisfying*

$$\mu > 1 \quad \text{and} \quad \mu \geq \frac{1}{d/e - 1},$$

every $c \in \mathrm{PCF}(d,e)$ satisfies the height bound

$$h(c/d) \leq \log\left((d-e)^{(d-e)/(d-1)}\mu^{(d-e)/(d-1)}\left(\frac{d/e}{1-1/\mu}\right)^{1/(d-1)}\right).$$

In particular, the set $\mathrm{PCF}(d,e)$ is a set of bounded height.

PROOF. An elementary, but lengthy (and currently unpublished), calculation. $\square$

6.3. The invariant measure and the Lyapunov exponent

In this section we work over $\mathbb{C}$. The distribution of the periodic points of a rational map are closely related to its dynamics. In particular, there are only finitely many nonrepelling periodic points, so almost all periodic points are in the Julia set, and indeed they are dense. The next theorem summarizes these and a few other of the many known results relating the periodic points to the Julia set.

Theorem 6.39. *Let $\phi \in \mathrm{Hom}_d(\mathbb{C})$ with $d \geq 2$.*

(a) *ϕ has at most $2d-2$ nonrepelling periodic points.*

(b) *$\mathrm{Per}(\phi)$ is dense the Julia set $\mathcal{J}(\phi)$, where we are working in the complex topology.*

(c) *For any $P \in \mathbb{P}^1(\mathbb{C})$, let μ_P denote the probability measure with a point mass at P. For each $n \geq 1$, define*

$$\mu_{\phi,n} = \frac{1}{d^n+1}\sum_{P \in \mathrm{Per}_n(\phi)} \mu_P$$

to be the discrete measure that is equidistributed on the points of period n, counted with appropriate multiplicities. Then

$$\hat{\mu}_\phi = \lim_{n\to\infty} \mu_{\phi,n}$$

converges in the weak-$*$ topology to a measure that is supported on the Julia set $\mathcal{J}(\phi)$.

(d) *The measure $\hat{\mu}_\phi$ satisfies*

$$\phi^*\hat{\mu}_\phi = d\hat{\mu}_\phi \quad \text{and} \quad \phi_*\hat{\mu}_\phi = \hat{\mu}_\phi.$$

Definition 6.40. The measure $\hat{\mu}_\phi$ defined in Theorem 6.39(c) is called the *invariant measure associated to* ϕ. In the literature it goes by many names, including the *equilibrium measure*, the *measure of maximal entropy*, the *Lyubich measure*, and the *Brolin–Lyubich measure.*

Remark 6.41. There is also an invariant measure in the p-adic setting, but it is a measure on Berkovich space, not on projective space. This is reasonable, since the space $\mathbb{P}^1(\mathbb{C}_p)$ is not locally compact, so its subsets do not naturally admit nice measures.

The magnitude of the derivative of a map at a point is a measure of the extent to which the map is expanding or contracting, so it is not surprising that the function

$$P \longrightarrow \bigl|\phi'(P)\bigr|$$

plays a key role in the dynamics of ϕ. However, the ordinary derivative with respect to z is too dependent on the choice of the coordinate z. A more intrinsic measure of expansion/contraction is the derivative relative to the chordal metric, as we now explain.

Definition 6.42. The *chordal metric* on $\mathbb{P}^1(\mathbb{C})$ is the function

$$\delta(P_1,P_2) = \delta\bigl([x_1,y_1],[x_2,y_2]\bigr) = \frac{|x_1y_2 - x_2y_1|}{\sqrt{|x_1^2|+|y_1^2|}\cdot\sqrt{|x_2^2|+|y_2^2|}}.$$

Proposition 6.43. *The chordal metric is a metric. More precisely, it satisfies:*

- $0 \le \delta(P_1,P_2) \le 1$ *for all* $P_1,P_2 \in \mathbb{P}^1(\mathbb{C})$.
- $\delta(P_1,P_2) = 0$ *if and only if* $P_1 = P_2$.
- $\delta(P_1,P_3) \le \delta(P_1,P_2) + \delta(P_2,P_3)$ *for all* $P_1,P_2,P_3 \in \mathbb{P}^1(\mathbb{C})$.

Definition 6.44. Let $\phi(z) \in \operatorname{Hom}_d(\mathbb{C})$ be a rational map, and let $P \in \mathbb{P}^1(\mathbb{C})$. The *chordal derivative of* ϕ *at* P is the quantity

$$\phi^{\#}(P) = \lim_{\substack{Q\in\mathbb{P}^1(\mathbb{C})\\ \delta(Q,P)\to 0}} \frac{\delta\bigl(\phi(Q),\phi(P)\bigr)}{\delta(Q,P)}. \tag{6.20}$$

(Note that the value of the chordal derivative is a nonnegative real number.)

The logarithmic average of the chordal derivative relative to the invariant measure is an important invariant of the map ϕ.

Definition 6.45. Let $\phi \in \operatorname{Hom}_d(\mathbb{C})$. The *Lyapunov exponent of* ϕ is the quantity

$$L(\phi) = \int_{\mathbb{P}^1(\mathbb{C})} \log\phi^{\#}(z)\,\hat{\mu}_\phi(z).$$

Remark 6.46. The Lyapunov exponent is PGL_2-conjugation invariant, so is a well-defined real-valued function on $\mathsf{M}_d(\mathbb{C})$.

As one might expect, the Lyapunov exponent of ϕ is related to both the critical points and the periodic points of ϕ, as in the following results.

Theorem 6.47. *Let* $\phi \in \operatorname{Hom}_d(\mathbb{C})$, *and let* $\hat{g}_\phi$ *be the associated Green function as defined in Section* 5.4.

(a) $$L(\phi) = \sum_{P\in\mathrm{Crit}(\phi)} \hat{g}_\phi(P) - \frac{2}{d}\log|\mathrm{Resultant}\,\phi| - \log d.$$

(Here Resultant ϕ *is the resultant of an appropriately normalized numerator and denominator of* ϕ*.)*

(b) $$L(\phi) = \lim_{n\to\infty} \frac{1}{d^n} \sum_{P\in\mathrm{Per}_n(\phi)} \log^+\bigl|\lambda_\phi(P)\bigr|.$$

(Here $\log^+(t) = \max\{\log(t), 0\}$*.)*

PROOF. See for example [30] or [92]. □

We can also work in the p-adic setting, say over $\mathbb{C}_p$.

Definition 6.48. The *p-adic chordal metric* on $\mathbb{P}^1(\mathbb{C}_p)$ is the function

$$\delta_p(P_1, P_2) = \delta_p\bigl([x_1, y_1], [x_2, y_2]\bigr) = \frac{|x_1y_2 - x_2y_1|_p}{\max\bigl\{|x_1|_p, |y_1|_p\bigr\} \cdot \max\bigl\{|x_2|_p, |y_2|_p\bigr\}}.$$

The p-adic chordal derivative of a map $\phi \in \mathrm{Hom}_d(\mathbb{C}_p)$ at a point $P \in \mathbb{P}^1(\mathbb{C}_p)$ is defined using the same formula (6.20) as is used when working over $\mathbb{C}$.

There is also a Lyapunov exponent in the non-Archimedean setting, although the definition uses Berkovich space, and there are p-adic versions of Theorem 6.47 in varying degrees of generality. See for example [92].

6.4. Conservative maps

The simplest sort of critical point portrait is one in which every critical point has a single arrow pointing to itself, i.e., every critical point is fixed. This prompts the following definition.

Definition 6.49. A rational map $\phi(z) \in \mathbb{C}(z)$ is *conservative* if all of its critical points are fixed points.

For a given degree, there are clearly only finitely many critical point portraits in which all of the critical points are fixed, so it follows from Thurston's rigidity theorem (Theorem 6.7) that there are only finitely many such maps (up to PGL_2 conjugation).

Example 6.50. An easy example of a conservative polynomial is obtained by setting $\phi(z) = z^d - dc^{d-1}z$ and choosing an appropriate value for c. The critical points are $\mathrm{Crit}(\phi) = \{\zeta c : \zeta \in \boldsymbol{\mu}_{d-1}\}$, and setting $\phi(\zeta c) = \zeta c$ yields the conservative polynomial

$$\phi(z) = z^d + \frac{d}{d-1}z.$$

Example 6.51. It is more difficult to produce nontrivial examples of nonpolynomial rational functions that are conservative. Here are examples, due to David Speyer [114], which show that such maps exist in every degree:

$$\phi(z) = \frac{(d-2)z^d + dz}{dz^{d-1} + (d-2)}.$$

The fixed points of $\phi(z)$ are 0, ∞, and the $(d-1)^{\mathrm{st}}$-roots of unity, while the critical points of $\phi(z)$ are the $(d-1)^{\mathrm{st}}$-roots of unity, each of which occurs with multiplicity 2. This can be seen from the elementary calculation

$$\phi(z) - z = \frac{-2z(z^{d-1}-1)}{dz^{d-1} + (d-2)} \quad\text{and}\quad \phi'(z) = \frac{d(d-2)(z^{d-1}-1)^2}{(dz^{d-1} + (d-2))^2}.$$

Notice that $\phi(z)$ is defined over $\mathbb{Q}$, as is the conservative polynomial in Example 6.50. It seems an interesting problem to classify all conservative maps in $\mathbb{Q}[z]$ or $\mathbb{Q}(z)$. In particular, is the number of conservative maps of degree d in $\mathbb{Q}[z]$ or $\mathbb{Q}(z)$ bounded independently of d? (Of course, we are really counting $\mathrm{PGL}_2(\mathbb{Q})$ equivalence classes of maps.)

Definition 6.52. A polynomial $\phi(z)$ is *normalized* if it is monic and $\phi(0) = 0$. Notice that over an algebraically closed field, every polynomial of degree $d \geq 2$ can be put into normalized form via conjugation by an affine transformation $f(z) = \alpha z + \beta$.

Proposition 6.53 (Tischler [118]). *Let $d \geq 2$. Then there are exactly $\binom{2d-2}{d-1}$ normalized conservative polynomials of degree d in $\mathbb{C}[z]$.*

The coefficients of conservative polynomials are in $\overline{\mathbb{Q}}$, and thus $\mathrm{Gal}(\overline{\mathbb{Q}}/\mathbb{Q})$ acts on the set of conservative polynomials of degree d. Pakovich [94] has studied this action in terms of an action on a certain tree.

Definition 6.54. The (absolute projective) height of a polynomial

$$\phi(z) = c\prod_{i=1}^{d}(z - \alpha_i) \in \overline{\mathbb{Q}}[z]$$

is defined to be

$$\frac{1}{d}\sum_{i=1}^{d} h(\alpha_i).$$

We note that if $\phi \in \mathbb{Q}[z]$ is irreducible over $\mathbb{Q}$ and if $\alpha \in \overline{\mathbb{Q}}$ is any root of ϕ, then $h(\phi)$ is equal to $h(\alpha)$.

Question 6.55. Let $\phi(z) = z^d + \big(d/(d-1)\big)z$ be the conservative polynomial from Example 6.50. Then

$$h(\phi) = \frac{1}{d}\log d,$$

so in particular, the height goes to 0 as $d \to \infty$. This raises a natural question. Let $\mathcal{C}_d^{\mathrm{poly}} \subset \overline{\mathbb{Q}}[x]$ denote the set of normalized conservative polynomials of degree d. Is it true that

$$\lim_{d\to\infty} \max_{\phi\in\mathcal{C}_d^{\mathrm{poly}}} h(\phi) = 0?$$

Extrapolating from the one conservative polynomial described in Example 6.50, one might even ask if

$$\limsup_{d\to\infty} \max_{\phi\in\mathcal{C}_d^{\mathrm{poly}}} \frac{h(\phi)}{(\log d)/d} < \infty?$$

6.5. A dynamical André–Oort conjecture

The rough equivalences in Figure 6.1 are often used in formulating analogies between arithmetic geometry and dynamical systems.

Thus if we let A_g denote the moduli space of principally polarized abelian varieties of dimension g, then the "special" points in A_g, namely the points corresponding to abelian varieties with complex multiplication, should be analogous in some sense to the "special" points in M_d, namely the points corresponding to PCF maps. The analogy is far from perfect, but at least in both cases the special points

Arithmetic Geometry		Dynamical Systems
rational and integral points on varieties	$\longleftrightarrow$	rational and integral points in orbits
torsion points on abelian varieties	$\longleftrightarrow$	periodic and preperiodic points of rational maps
abelian varieties with CM	$\longleftrightarrow$	postcritically finite rational maps

FIGURE 6.1

are naturally partitioned into a countable union of finite sets. For PCF maps, each finite set is determined by a critical point portrait, while for CM abelian varieties, each finite set is determined by an order in a CM field. The classical André–Oort conjecture [1,93] says roughly that if an irreducible subvariety $X \subset (\mathsf{A}_g)^n$ contains a Zariski closed set of special, i.e., CM, points, then it is essentially a subvariety associated to a Hecke correspondence. Jonathan Pila [97] has proven the classical André–Oort conjecture for products of elliptic modular curves using methods from model theory, O-minimal geometry, and transcendence theory.

It is natural to make an analogous conjecture for dynamical systems, but it is far from clear how to formulate the conclusion. We recall from Definition 6.8 that $\mathsf{M}_d^{\text{crit}}$ is the moduli space of rational maps of degree d with marked critical points.

Conjecture 6.56 (Dynamical André–Oort Conjecture; Baker–DeMarco, personal communication). *Let $X \subset \mathsf{M}_d^{\text{crit}}$ be an algebraic subvariety such that the PCF maps in X are Zariski dense in X. Then X is cut out by critical orbit relations.*

The condition of being "cut out by critical orbit relations" is meant to describe the dynamical condition that there are at most $\dim(X)$ "free" critical orbits. The simplest sorts of critical orbit relations are those of the form

$$\phi^n(c_i) = \phi^m(c_j),$$

but this does not take into account possible symmetries. For example, if $f \in \operatorname{Aut}(\phi)$, then an equation of the form

$$\phi^n(c_i) = f \circ \phi^m(c_j) \tag{6.21}$$

would also be a critical orbit relation.

However, these are not the only "extra" critical point relations. Ingram has pointed out that critical point relations coming from lower degree iterates can cause problems in any naive attempt to give a precise statement of a dynamical André–Oort conjecture. Ingram uses the following family of maps to illustrate what can go wrong.

Proposition 6.57. *Let*

$$\phi_t(z) = z^2 - t^2 \quad \textit{and} \quad \psi_t(z) = \phi_t^2(z) = (z^2 - t^2)^2 - t^2.$$

The finite critical points of ψ_t are $0, t$ and $-t$, and the only stable critical point relation between these three points is the obvious equation

$$\psi_t(t) = \psi_t(-t).$$

(*Here stable means that the critical point relation persists in a neighborhood of* t.)

PROOF. The finite critical points of ψ_t can be read off from the formula

$$\psi_t'(z) = 4z(z^2 - t^2) = 0.$$

The relation $\psi_t(t) = \psi_t(-t)$ is obvious. Suppose that there is another critical point relation. It must involve 0 and be of the form

$$\psi_t^n(\pm t) = \psi_t^m(0).$$

If $n = 0$, we apply ψ_t to both sides, which allows us to assume that $n \geq 1$. Since $\psi_t = \phi_t^2$ and $\phi_t(\pm t) = 0$, we get

$$\phi_t^{2n-1}(0) = \phi_t^{2n}(\pm t) = \phi_t^{2m}(0).$$

But clearly $2n - 1 \neq 2m$, so this gives a stable critical point relation for the family of maps $z^2 - t^2$, which is impossible. □

Remark 6.58 (Ingram, personal communication). Consider the curve in M_4 defined by the polynomial maps $\psi_t(z) = (z^2 - t^2)^2 - t^2$ in Proposition 6.57,

$$\mathcal{C} = \bigl\{\langle\psi_t\rangle \in \mathsf{M}_4 : t \in \mathbb{C}\bigr\}.$$

Clearly $\mathcal{C}$ is a subset of the set of polynomial maps $\mathcal{P}_4 \subset \mathsf{M}_4$, where we note that $\mathcal{P}_4$ is defined by a critical point relation, namely it consists of all maps having a fixed critical point of degree 4. The space $\mathcal{P}_4$ has dimension 3, and $\mathcal{C}$ is a curve in $\mathcal{P}_4$.

The PCF points in $\mathcal{C}$ are Zariski dense in $\mathcal{C}$, because ψ_t is PCF if and only if $\phi_t = z^2 - t^2$ is PCF, and we know that there are infinitely many PCF maps of the form ϕ_t. On the other hand, Proposition 6.57 tells us that as a subvariety of $\mathcal{P}_4$, there is only one stable critical point relation satisfied by the maps ψ_t that define the curve $\mathcal{C}$. So the smallest subvariety of M_4 defined by critical point conditions and containing $\mathcal{C}$ is a two-dimensional hypersurface $\mathcal{V}$ in $\mathcal{P}_4$. (It is interesting to note that $\mathcal{V}$ seems to be reducible and not reduced.) Thus $\mathcal{C}$ has a dense set of PCF points, yet is not cut out by naive critical point relations of the form (6.21).

The problem is that there is a hidden critical point relation for the family of maps ψ_t that can be seen from the family ϕ_t, but is not directly visible from ψ_t. This example illustrates some of the subtleties implicit in formulating a precise dynamical André–Oort conjecture.

In a positive direction, DeMarco has noted that a result of Ghioca, Hsia, and Tucker [43] (cf. Remark 5.26) may be viewed as a special case of Conjecture 6.56 for polynomials (with some extra technical conditions). DeMarco has also made the following intriguing conjecture.

Conjecture 6.59 (DeMarco, private communication). *For each integer* $n \geq 1$ *and* $\lambda \in \mathbb{C}$, *let* $\operatorname{Per}_n(\lambda) \subset \mathsf{M}_2 \cong \mathbb{A}^2$ *denote the set of rational maps that have a periodic point of period* n *and multiplier* λ. *(It is not hard to see that* $\operatorname{Per}_n(\lambda)$ *is a curve in* M_2.*) Then* $\operatorname{Per}_n(\lambda)$ *contains infinitely many PCF maps if and only if* $\lambda = 0$.

DeMarco remarks that standard dynamical methods show that $\operatorname{Per}_n(\lambda)$ contains no PCF maps for $0 < |\lambda| < 1$, while for $|\lambda| > 1$, it is expected that there are two free critical orbits.

CHAPTER 7

Field of Moduli and Field of Definition

For details and generalizations of the results in this chapter, see [107].

7.1. Twists, automorphisms, and cohomology

Example 7.1. Two elliptic curves E and E' are isomorphic if and only if

$$j(E) = j(E'). \quad \textbf{True or False?}$$

The answer is that this statement is true when working over an algebraically closed field, but it is not true in general. For example, all of the curves

$$E_B : y^2 = x^3 + B$$

are isomorphic over $\overline{\mathbb{Q}}$, but $E_B \cong E_{B'}$ over $\mathbb{Q}$ if and only if $B = B'd^6$ for some $d \in \mathbb{Q}^*$. This distinction is very important, because the arithmetic properties of E depend on its $\mathbb{Q}$-isomorphism class. For example, the Mordell–Weil group $E_B(\mathbb{Q})$ varies for different values of $B \in \mathbb{Q}^*$.

Definition 7.2. We say that two maps $\phi, \psi \in \mathrm{Rat}_d^n(K)$ are *K-equivalent* (or *K-isomorphic*) if

$$\psi = \phi^f \quad \text{for some } f \in \mathrm{PGL}_{n+1}(K).$$

The dynamics of K-equivalent maps are identical in their action on the set of K-rational points $\mathbb{P}^n(K)$, but maps that are only $\overline{K}$-equivalent may behave quite differently on $\mathbb{P}^n(K)$. This is analogous to the fact that even if two elliptic curves E_1/K and E_2/K are $\overline{K}$-isomorphic, their Mordell–Weil groups $E_1(K)$ and $E_2(K)$ need not be the same.

Definition 7.3. Let $\phi \in \mathrm{Rat}_d^n(K)$. A *$K$-twist of ϕ* is a map $\psi \in \mathrm{Rat}_d^n(K)$ that is $\overline{K}$-equivalent to ϕ. We write

$$\mathrm{Twist}_K(\phi) = \frac{\{K\text{-twists of } \phi\}}{K\text{-equivalence}}.$$

Example 7.4. The maps

$$\phi(z) = z + \frac{1}{z} \quad \text{and} \quad \psi(z) = z - \frac{1}{z}$$

are $\mathbb{R}$-twists of one another, since

$$\psi(z) = i^{-1}\phi(iz) = \phi^f(z), \quad \text{where } f(z) = iz.$$

But they are not $\mathbb{R}$-equivalent, since for example

$$\mathrm{Per}_2^*(\phi) = \pm\sqrt{-\tfrac{1}{2}} \quad \text{and} \quad \mathrm{Per}_2^*(\psi) = \pm\sqrt{\tfrac{1}{2}},$$

while an $\mathbb{R}$-equivalence would identify $\mathrm{Per}_2^*(\phi) \cap \mathbb{R}$ with $\mathrm{Per}_2^*(\psi) \cap \mathbb{R}$.

Let

$$\langle\cdot\rangle\colon \operatorname{Hom}_d^n \longrightarrow \mathsf{M}_d^n$$

denote as usual the natural projection map from Hom_d^n to its quotient. Then for $\phi,\psi\in\operatorname{Hom}_d^n(K)$ we have

$$(\psi \text{ is a } K\text{-twist of } \phi) \iff \langle\phi\rangle = \langle\psi\rangle.$$

So the study of twists may be viewed as the study of the fibers of the map

$$\langle\cdot\rangle\colon \operatorname{Hom}_d^n(K)/\operatorname{PGL}_{n+1}(K) \longrightarrow \mathsf{M}_d^n(K).$$

We recall from Section 2.6 that the *automorphism group* of a map $\phi\in\operatorname{Rat}_d^n$ is the group

$$\operatorname{Aut}(\phi) = \{f\in\operatorname{PGL}_{n+1} : \phi^f = \phi\}.$$

As we will see, the size of the automorphism group helps determine the number of twists of ϕ.

Example 7.5. The map

$$\phi(z) = az + \frac{b}{z} \text{ satisfies } \phi(z) = -\phi(-z), \text{ so } f(z) = -z \text{ is in } \operatorname{Aut}(\phi).$$

Example 7.6. One can check that the map

$$\phi(z) = \frac{z^2-2z}{-2z+1} \text{ has } \operatorname{Aut}(\phi) = \left\{z, \frac{1}{z}, \frac{z-1}{z}, \frac{1}{1-z}, \frac{z}{z-1}, 1-z\right\} \cong \mathcal{S}_3.$$

Elliptic curves always have an automorphism $P\to -P$, but as we saw earlier (Corollary 2.55), most rational maps have no automorphisms.

We will use Galois theory to study the twists of maps. For ease of exposition, we assume for the rest of this section that

$$\boxed{K \text{ is a perfect field,}}$$

for example a field of characteristic zero. We let an element $\sigma\in\operatorname{Gal}(\overline{K}/K)$ act on a rational map $\phi\in\operatorname{Rat}_d^n(\overline{K})$ by applying σ to each of the coefficients of the polynomials defining ϕ.

Proposition 7.7. *Let $\phi\in\operatorname{Hom}_d^n(K)$. Then*

$$\operatorname{Aut}(\phi) = 1 \implies \#\operatorname{Twist}_K(\phi) = 1.$$

PROOF. Suppose first that $\operatorname{Aut}(\phi)=1$. Let $\psi\in\operatorname{Twist}_K(\phi)$, so there is an $f\in\operatorname{PGL}_{n+1}(\overline{K})$ such that

$$\psi = \phi^f = f^{-1}\circ\phi\circ f.$$

Applying $\sigma\in\operatorname{Gal}(\overline{K}/K)$ gives

$$\begin{aligned}\phi^f &= \psi = \sigma(\psi) = \sigma(\phi^f)\\ &= \sigma(f^{-1}\circ\phi\circ f) = \sigma(f^{-1})\circ\sigma(\phi)\circ\sigma(f) = \sigma(f^{-1})\circ\phi\circ\sigma(f) = \phi^{\sigma(f)}.\end{aligned}$$

Thus $\sigma(f)\circ f^{-1}\in\operatorname{Aut}(\phi)$, so $\sigma(f)=f$. Therefore $f\in\operatorname{PGL}_{n+1}(K)$, so ψ is K-equivalent to ϕ. □

Remark 7.8. It follows from Hilbert's Theorem 90 that

$$\operatorname{PGL}_{n+1}(K) = \operatorname{GL}_{n+1}(K)/K^*.$$

The proof of Proposition 7.7 also indicates how a nontrivial automorphism group can lead to nontrivial twists.

Quick Review of Group Cohomology. Let G be a group that acts on another (possibly noncommutative) group A. A 1-*cocycle* is a map

$$g\colon G \longrightarrow A \text{ satisfying } g_{\sigma\tau} = g_\sigma \cdot \sigma(g_\tau) \text{ for all } \sigma, \tau \in G.$$

Two 1-cocycles g and h are *cohomologous* if there is an element $f \in A$ such that

$$h_\sigma = f \cdot g_\sigma \cdot \sigma(f^{-1}) \quad \text{for all } \sigma \in G.$$

(It is not hard to check that this is an equivalence relation. Also, if A is abelian, this is the same as saying that $(h \cdot g^{-1})_\sigma = f \cdot \sigma(f^{-1})$ is a 1-coboundary as usual.) The *cohomology set* $H^1(G, A)$ is the set

$$H^1(G,A) = \frac{\{\text{1-cocycles } G \to A\}}{(g \sim h \text{ if } g \text{ and } h \text{ are cohomologous})}.$$

If the group A is abelian, then $H^1(G, A)$ is a group in the natural way,

$$(g \cdot h)_\sigma = g_\sigma \cdot h_\sigma,$$

but in general $H^1(G, A)$ is only a pointed set.

Example 7.9. Let $\boldsymbol{\mu}_n \subset \overline{\mathbb{Q}}^*$ be the n'th roots of unity. Then for any number field $K/\mathbb{Q}$,

$$H^1\big(\operatorname{Gal}(\overline{K}/K), \boldsymbol{\mu}_n\big) \cong K^*/K^{*n}.$$

This comes from the Kummer sequence

$$\begin{array}{ccccccc}
H^0(\overline{K}^*) & \xrightarrow{z \to z^n} & H^0(\overline{K}^*) & \longrightarrow & H^1(\boldsymbol{\mu}_n) & \longrightarrow & H^1(\overline{K}^*) \\
\| & & \| & & \| & & \|\ \text{Hilbert theorem 90} \\
K^* & \xrightarrow{z \to z^n} & K^* & \longrightarrow & H^1(\boldsymbol{\mu}_n) & \longrightarrow & 0.
\end{array}$$

Proposition 7.10. *Fix $\phi \in \operatorname{Rat}_d^n(K)$, and let ψ be a K-twist of ϕ, say*

$$\psi = \phi^f \text{ with } f \in \operatorname{PGL}_{n+1}(\overline{K}).$$

(a) *For all $\sigma \in \operatorname{Gal}(\overline{K}/K)$,*

$$f \circ \sigma(f^{-1}) \in \operatorname{Aut}(\phi).$$

(b) *The map*

$$g\colon \operatorname{Gal}(\overline{K}/K) \longrightarrow \operatorname{Aut}(\phi), \qquad g_\sigma = f \circ \sigma(f^{-1}),$$

is a 1-cocycle.

(c) *ψ is K-equivalent to ϕ if and only if g is a 1-coboundary.*

PROOF. This is all just unsorting of definitions. We essentially proved (a) during the course of proving Proposition 7.7. As a further illustration, we'll prove one direction of (c).

g is a 1-coboundary

$$\begin{aligned}
&\iff g_\sigma = h \circ \sigma(h^{-1}) \quad \text{for some } h \in \operatorname{Aut}(\phi), \text{ all } \sigma \in \operatorname{Gal}(\overline{K}/K),\\
&\iff f \circ \sigma(f^{-1}) = h \circ \sigma(h^{-1})\\
&\implies h^{-1} \circ f = \sigma(h^{-1} \circ f)\\
&\implies h^{-1} \circ f \in \operatorname{PGL}_{n+1}(K)\\
&\implies \phi \sim_{/K} \phi^{h^{-1}f} = \phi^f = \psi.
\end{aligned}$$

□

Thus twists of ϕ lead to 1-cocycles, but it's not quite true that every 1-cocycle gives a twist.

Theorem 7.11. *Fix $\phi \in \mathrm{Rat}_d^n(K)$ and let*

$$g\colon \mathrm{Gal}(\overline{K}/K) \longrightarrow \mathrm{Aut}(\phi)$$

be a 1-cocycle. The following are equivalent:

(i) *There is a K-twist of ϕ whose 1-cocycle is g.*
(ii) *g becomes a 1-coboundary when extended to*

$$g\colon \mathrm{Gal}(\overline{K}/K) \longrightarrow \mathrm{PGL}_{n+1}(\overline{K}).$$

This gives the following cohomological description of the twists of a map:

$$\mathrm{Twist}_K(\phi) = \left\{ \xi \in H^1\bigl(G_K, \mathrm{Aut}(\phi)\bigr) : \begin{matrix} \xi \text{ becomes trivial in} \\ H^1\bigl(G_K, \mathrm{PGL}_{n+1}(\overline{K})\bigr) \end{matrix} \right\}.$$

7.2. Fields of definition and field of moduli

In this section we consider the action of $\mathrm{Gal}(\overline{K}/K)$ on the coefficients of a map $\phi \in \mathrm{Rat}_d^n(\overline{K})$, and we study the question of whether there exists a change of variables such that ϕ^f has coefficients are in K.

Example 7.12. The map

$$\phi(z) = z^3 + i$$

is defined over $\mathbb{Q}(i)$, but the change of variables $f(z) = iz$ gives a map

$$\phi^f(z) = f^{-1} \circ \phi \circ f(z) = -i\phi(iz) = -z^3 + 1$$

that is defined over $\mathbb{Q}$.

Definition 7.13. Let $\phi(z) \in \mathrm{Rat}_d^n(\overline{K})$. A field K'/K is a *field of definition for ϕ* (FOD) if there exists some $f \in \mathrm{PGL}_{n+1}(\overline{K})$ such that

$$\phi^f \in \mathrm{Rat}_d^n(K').$$

Note that a map ϕ will have many fields of definition.

Definition 7.14. Let $\phi(z) \in \mathrm{Rat}_d^n(\overline{K})$, and let

$$G_\phi = \bigl\{\sigma \in \mathrm{Gal}(\overline{K}/K) : \sigma(\phi) \text{ is } \overline{K}\text{-equivalent to } \phi\bigr\}.$$

The *field of moduli for ϕ* (FOM) is the fixed field of G_ϕ,

$$K_\phi = \overline{K}^{G_\phi}.$$

Note that the field of moduli of ϕ is a specific field, determined by ϕ.

It is easy to see that every field of definition contains the field of moduli, and it is not hard to show that the field of moduli is the intersection of all fields of definition. In particular, if

$$\mathrm{FOM} = \mathrm{FOD},$$

then there is a minimal field of definition for ϕ, namely the field of moduli K_ϕ.

Remark 7.15. The name *field of moduli* reflects the fact that

$$K_\phi = K\big(\langle\phi\rangle\big),$$

where $\langle\phi\rangle \in \mathsf{M}_d^n(\overline{K})$ is the moduli point associated to ϕ. (This is clear from the definitions.) In particular, the maps satisfying FOM = FOD are exactly the maps in the image of the natural projection map

$$\langle\cdot\rangle\colon \operatorname{Hom}_d^n(K) \longrightarrow \mathsf{M}_d^n(K).$$

The reason that this map need not be surjective is because $\mathsf{M}_d^n(K)$ need not be equal to $\operatorname{Hom}_d^n(K)/\operatorname{PGL}_{n+1}(K)$. Instead, we have

$$\mathsf{M}_d^n(K) = \left(\frac{\operatorname{Hom}_d^n(\overline{K})}{\operatorname{PGL}_{n+1}(\overline{K})}\right)^{\operatorname{Gal}(\overline{K}/K)}.$$

Example 7.16. Let

$$\phi(z) = i\left(\frac{z-1}{z+1}\right)^3. \tag{7.1}$$

Clearly $\mathbb{Q}(i)$ is a FOD for ϕ. Letting τ be complex conjugation, it is easy to verify that

$$\tau(\phi)(z) = -i\left(\frac{z-1}{z+1}\right)^3 = \phi^g(z) \quad \text{with } g(z) = -1/z.$$

Hence the FOM of ϕ is $\mathbb{Q}$. On the other hand, it's also not hard to check that

$$\phi^f(z) \notin \mathbb{Q}(z) \quad \text{for all } f \in \operatorname{PGL}_2(\overline{\mathbb{Q}}),$$

so $\mathbb{Q}$ is *not* a FOD for ϕ.

The map (7.1) has an interesting property, namely that $\phi^3(z) \in \mathbb{Q}(z)$. This property is uncommon even among maps whose FOM is $\mathbb{Q}$, but for which $\mathbb{Q}$ is not a FOD. (The observation that $\phi^3(z) \in \mathbb{Q}(z)$ is due to Mike Zieve.)

7.3. Tools for determining when FOM = FOD

For this section we fix some $\phi \in \operatorname{Rat}_d^n(\overline{K})$ and take

$$K = K_\phi = (\text{the FOM for } \phi).$$

Then $\sigma(\phi)$ is equivalent to ϕ for every $\sigma \in \operatorname{Gal}(\overline{K}/K)$, so

$$\sigma(\phi) = \phi^{g_\sigma} \quad \text{for some } g_\sigma \in \operatorname{PGL}_{n+1}(\overline{K}).$$

To simplify the exposition, we will also assume that

$$\boxed{\operatorname{Aut}(\phi) = 1.}$$

This ensures that g_σ is uniquely determined by σ, so we obtain a well-defined map

$$\operatorname{Gal}(\overline{K}/K) \longrightarrow \operatorname{PGL}_{n+1}(\overline{K}), \quad \sigma \longmapsto g_\sigma. \tag{7.2}$$

Proposition 7.17. *Let $\phi \in \operatorname{Rat}_d^n(\overline{K})$ be a map with $\operatorname{Aut}(\phi) = 1$.*

(a) *The map (7.2) is a 1-cocycle.*
(b) *K is a FOD for ϕ if and only if the map (7.2) is a 1-coboundary.*

PROOF. Unsorting definitions. □

Remark 7.18. In general, for any perfect field K and integer $n \geq 0$, there is a bijection of sets

$$H^1\big(G_K, \mathrm{PGL}_{n+1}(\overline{K})\big) \xrightarrow{\sim} H^2(G_K, \overline{K}^*)[n+1] = \mathrm{Br}(K)[n+1]. \tag{7.3}$$

(Here $\mathrm{Br}(K)[n+1]$ is the $(n+1)$-torsion subgroup of the Brauer group of K.) One can often use (7.3) to determine if ϕ satisfies FOM = FOD.

The map (7.3) comes from the long exact sequence associated to

$$1 \longrightarrow \overline{K}^* \longrightarrow \mathrm{GL}_{n+1}(\overline{K}) \longrightarrow \mathrm{PGL}_{n+1}(\overline{K}) \longrightarrow 1,$$

and the "injectivity" of (7.3) follows from Hilbert's theorem 90 for GL_n. However, since $H^1\big(G_K, \mathrm{PGL}_{n+1}(\overline{K})\big)$ is only a pointed set, general principles only imply that the map (7.3) is an injective map of pointed sets. By definition, this means that the marked point in $H^1(\mathrm{PGL})$ is the only element that maps to the identity element in $\mathrm{Br}(K)$, but it does not imply that the map (7.3) is an injection of sets. See [100, §X.5] for a proof that (7.3) is a bijection, and see also [100, Appendix to Chapter VII] for a discussion as to why the injectivity of (7.3) is not an automatic consequence of the fact that $H^1\big(G_K, \mathrm{GL}_{n+1}(\overline{K})\big) = 0$.

The following result is for $n = 1$. The theorem does not require the assumption that $\mathrm{Aut}(\phi) = 1$, but the proof is considerably more difficult without that assumption.

Theorem 7.19 (Silverman [107]). *Let $\phi \in \mathrm{Rat}_d^1(\overline{K})$ with $d \geq 2$. Then*

$$\mathrm{FOM} = \mathrm{FOD}$$

in the following two cases:

(a) *d is even.*

(b) *ϕ is a polynomial map, i.e., $\phi(z)$ is conjugate to a map in $\overline{K}[z]$.*

PROOF SKETCH FOR $\mathrm{Aut}(\phi) = 1$. The map ϕ gives us a 1-cocycle

$$g_\phi\colon \mathrm{Gal}(\overline{K}/K) \to \mathrm{PGL}_2(\overline{K}).$$

The elements of $H^1\big(G_K, \mathrm{PGL}_2(\overline{K})\big)$ classify twists of $\mathbb{P}^1$, i.e., they classify algebraic varieties X/K that are isomorphic to $\mathbb{P}^1$ over $\overline{K}$. (These twists are called homogeneous spaces.) Let X_ϕ/K be the twist of $\mathbb{P}^1$ corresponding to the cocycle g_ϕ. Then

$$\begin{aligned} X_\phi \text{ is the trivial twist} &\iff X_\phi(K) \neq \varnothing \\ &\iff g_\phi \text{ is a 1-coboundary.} \end{aligned}$$

From general principles, any twist of $\mathbb{P}^1$ becomes trivial in a quadratic extension. (In general, a twist of $\mathbb{P}^n$ becomes trivial in an extension of degree $n+1$.)

The map $\phi\colon \mathbb{P}^1 \to \mathbb{P}^1$ is only defined over $\overline{K}$, but an easy computation shows that the map $g_\phi^{-1} \circ \phi \circ g_\phi$ induces a map $\psi\colon X_\phi \to X_\phi$ that is defined over K. Suppose now that d is even. Then the sum of the fixed points of ψ gives a divisor of degree $d+1$ on X_ϕ that is defined over K. Since d is even, this divisor has odd degree, and since we also have a divisor of degree 2 defined over K, we can get a divisor of degree 1 on X_ϕ that is defined over K. Then the Riemann–Roch theorem gives a point in $X_\phi(K)$, so X_ϕ is a trivial homogeneous space.

Similarly, if ϕ is a polynomial map, then ψ has a unique totally ramified fixed point (unless ϕ is conjugate to z^d), so this point must be fixed by $\mathrm{Gal}(\overline{K}/K)$, again showing that $X_\phi(K) \neq \varnothing$.

For further details, and the case-by-case analysis that is required when $\mathrm{Aut}(\phi) \neq 1$, see [107]. $\square$

Remark 7.20. Hutz and Manes (in preparation) have partially generalized Theorem 7.19(a). They prove that if $\phi\colon \mathbb{P}^N \to \mathbb{P}^N$ is a morphism of degree $d \geq 2$ with $\mathrm{Aut}(\phi) = 1$, then

$$\gcd(d^N + 1, N + 1) = 1 \implies \mathrm{FOM} = \mathrm{FOD}\,.$$

Schedule of Talks at the Bellairs Workshop

- **Monday:**
  - Joseph Silverman: *Introduction to Moduli Spaces*
  - Laura DeMarco: *Moduli Spaces of One-Dimensional Dynamical Systems*
- **Tuesday:**
  - Joseph Silverman: *Dynamical Moduli Spaces*
  - Adam Epstein: *Transversality and Holomorphic Dynamics*
- **Wednesday:**
  - Joseph Silverman: *Dynatomic Polynomials and Dynamical Modular Curves*
  - Michelle Manes: *Level Structures on Dynamical Moduli Spaces*
- **Thursday:**
  - Joseph Silverman: *Canonical Heights*
    - Anupam Bhatnagar: *Points of Canonical Height Zero on Projective Varieties*
    - ChongGyu (Joey) Lee: *Height Estimates for Dominant Morphisms*
    - Alon Levy: *Semistable Reduction for Dynamical Systems*
- **Friday:**
  - Joseph Silverman: *Field of Moduli and Fields of Definition*
  - Michael Tepper: *Isotriviality and* M_d^n

Glossary

$[F_n, G_n]$	nth iterate of $[F, G]$, 57
$[\phi, \psi]_{\mathrm{AZ}}$	Arakelov–Zhang pairing, 72
$[\phi, \psi]_{\mathrm{KS}}$	Kawaguchi–Silverman pairing, 72
$F \asymp G$	notation for $F \ll G$ and $G \ll F$, 101
$g \cdot x$	the image of the action of g on x, 12
$\langle \phi \rangle$	the point in M_d^n corresponding to $\phi \in \mathrm{Hom}_d^n$, 25
$G \ll F$	notation for $G(x) \le c_1 F(x) + c_2$, 101
A^G	the ring of invariant of G acting on A, 11
A_g	the moduli space of principally polarized abelian varieties, 88
$\bar{\mathfrak{a}}(\phi)$	ideal class associated to minimal model of ϕ, 49
$\mathrm{Aut}(\phi)$	the automorphism group of the map ϕ, 34
BiCrit_d	set of maps with exactly two critical points, 42
$C_1(n)$	dynamical modular curve for maps $\psi_b(z) = z/(z^2 + b)$, 66
$\deg_X(S)$	minimal degree of polynomial vanishing on S, 98
$e_\phi(P)$	the ramification index of ϕ at P, 91
$e_v(\phi)$	the valuation of the Macaulay resultant of ϕ, 47
$\varepsilon_v(\phi)$	exponent at v of the minimal resultant of ϕ, 48
F_g	moduli space of K3 surfaces, 52
Fix	map from Hom_d to the fixed points of the map, 36
$\Phi_{\phi,n}$	n-dynatomic polynomial, 57
ϕ^f	$= f^{-1} \circ \phi \circ f$, the conjugate of ϕ by f, 4
ϕ^n	nth iterate of the function ϕ, 3
$\phi^\#(P)$	chordal derivative of ϕ at P, 108
$\psi_c(z)$	the polynomial $z^2 + c$, 59
G_ϕ	the subgroup of $\mathrm{Gal}(\overline{K}/K)$ such that $\sigma(\phi) \sim \phi$, 116
$\mathbb{G}_m$	the multiplicative group, 15
$\widehat{G}_{\Phi,v}(P)$	Green function for lift Φ of ϕ, 76
$\hat{g}_{\phi,v}$	Green function (local canonical height), 76
$H^1(G, A)$	cohomology set for G acting on A, 115
$\hat{h}_{\mathrm{crit}}(\phi)$	critical canonical height of the map ϕ, 99

$\hat{h}^+_\phi, \hat{h}^-_\phi$	canonical heights for a regular affine automorphism, 83
$\hat{h}^+_\phi, \hat{h}^-_\phi$	canonical heights on a K3 surface, 86
h	the Weil height on $\mathbb{P}^N(\bar{K})$, 69
$h(\phi)$	the height of the map ϕ via $\phi \in \mathrm{Hom}_d^N \subset \mathbb{P}^N$, 70
$(\mathrm{Hom}_d)^{\mathrm{ss}}$	semistable locus in Hom_d, 21
$(\mathrm{Hom}_d)^{\mathrm{s}}$	stable locus in Hom_d, 21
$\mathrm{Hom}_d(\mathbb{P}^1(\mathbb{C}))$	degree d rational self-maps of $\mathbb{P}^1(\mathbb{C})$, 3
Hom_d^n	degree d morphisms $\mathbb{P}^n \to \mathbb{P}^n$, 7
$\mathrm{Hom}_d^n(m)$	maps with a marked point of formal period n, 8
ι_1, ι_2	involutions on the surface $S_{\boldsymbol{A},\boldsymbol{B}}$, 51
$\mathcal{J}^{\mathrm{f}}(\phi)$	the filled Julia set of ϕ, 76
K_ϕ	the field of moduli for ϕ, 116
$L(\phi)$	the Lyapunov exponent of ϕ, 108
L_d	the set of flexible Lattès maps in M_d, 101
$\ell(0) \cdot x$	specialization of the 1-parameter subgroup ℓ, 17
Λ_ϕ^n	the multiplier spectrum of ϕ, 25
$\lambda_\phi(\alpha)$	multiplier of ϕ at the periodic point α, 24
Lat_d	the set of flexible Lattès maps in Hom_d, 101
$(\mathsf{M}_d)^{\mathrm{ss}}$	moduli space of semistable points in Hom_d, 21
$(\mathsf{M}_d)^{\mathrm{s}}$	moduli space of stable points in Hom_d, 21
M_K	complete set of inequivalent normalized absolute values on K, 69
M_K	set of absolute values on the function field $K = k(C)$, 45
M_d	moduli space of self-morphisms of $\mathbb{P}^1$, 8
M_d^n	moduli space of self-morphisms of $\mathbb{P}^n$, 8
$\mathsf{M}_d^n(m)$	moduli space of maps with a marked point of formal period n, 8
$\mathsf{M}_d^{\mathrm{BiCrit}}$	image of BiCrit_d in M_d, 43
$\mathsf{M}_d^{\mathrm{crit}}$	moduli space of degree d maps with marked critical points, 92
$\mathsf{M}_d^{\mathrm{crit}}[i](r,n)$	subvariety of $\mathsf{M}_d^{\mathrm{crit}}$ with marked critical point having specified portrait, 93
$\mathfrak{M}$	the Mandelbrot set, 88
$\overline{\mathsf{M}}_2$	completion of M_2, 33
$\hat{\mu}_\phi$	invariant measure associated to ϕ, 107
$\mu(\phi)$	height expansion coefficient of ϕ, 81
$\mu^{\mathcal{L}}(x,\ell)$	integer invariant attached to 1-parameter subgroup ℓ, 18
$\mathcal{O}_\phi(x)$	the forward orbit of x for the map ϕ, 3
$\mathcal{O}_{\phi,\psi}(P)$	full orbit of P for two maps ϕ and ψ, 80
$\mathsf{P}_d^{\mathrm{crit}}$	subvariety of $\mathsf{M}_d^{\mathrm{crit}}$ corresponding to polynomial maps, 96

$\mathsf{P}_d^{\mathrm{crit}}[i](r,n)$	subvariety of $\mathsf{P}_d^{\mathrm{crit}}$ with point having given critical point orbit, 96
$\mathrm{Per}(\phi, X)$	set of periodic points for ϕ, 4
$\mathrm{Per}_n(\phi, X)$	set of periodic points of period n for ϕ, 4
$\mathrm{Per}_n^*(\phi)$	periodic points of formal period n, 92
$\mathrm{Per}_n^{**}(\phi)$	points of exact period n, 8
PGL_{n+1}	the projective linear group, 8
$\mathrm{PrePer}(\phi, X)$	set of preperiodic points for ϕ, 4
$\mathrm{PrePer}_{m,n}(\phi, X)$	set of preperiodic points of tail m and period n, 4
$\mathrm{PrePer}_{r,n}^*(\phi)$	preperiodic points of tail length r and formal period n, 92
$\mathfrak{R}(\phi)$	the minimal resultant of ϕ, 48
Rat_d^n	degree d rational maps $\mathbb{P}^n \to \mathbb{P}^n$, 7
$(\mathcal{S}\phi)(z)$	the Schwarzian derivative of ϕ, 29
$S_{\boldsymbol{A},\boldsymbol{B}}$	K3 surface determined by the coefficients $\boldsymbol{A}$ and $\boldsymbol{B}$, 50
$\mathcal{S}_n$	symmetric group, 11
$\sigma(g, x)$	the image of the action of g on x, 12
$\sigma_{i,n}(\phi)$	symmetric function of multipliers of ϕ, 25
$\mathrm{Stab}(f)$	the stabilizer of the map f, 12
$\mathcal{T}_P(\mathbb{P}^1)$	tangent space of $\mathbb{P}^1$ at P, 25
$\mathcal{T}_P(\mathbb{P}^N)$	tangent space of $\mathbb{P}^N$ at P, 27
$\mathrm{Twist}_K(\phi)$	the set of K-twists of ϕ, 113
V/G	the quotient of V by the finite group G, 11
$X^{\mathrm{ss}}(\mathcal{L})$	semistable locus, 16
$X^{\mathrm{s}}(\mathcal{L})$	stable locus, 16
$X_0(n)$	smooth projective model for $Y_0(n)$, 60
$X_1(n)$	smooth projective model for $Y_1(n)$, 59
$X_{(0)}^{\mathrm{s}}(\mathcal{L})$	stable locus with dimension 0 stabilizer, 16
$Y_0(n)$	dynamical modular curve, 60
$Y_1(n)$	dynamical modular curve, 59
$Z(\phi)$	locus of indeterminacy of ϕ, 40

Bibliography

1. Y. André, *G-functions and geometry*, Aspects Math., vol. E13, Friedr. Vieweg, Braunschweig, 1989.
2. M. Baker, *A finiteness theorem for canonical heights attached to rational maps over function fields*, J. Reine Angew. Math. **626** (2009), 205–233.
3. M. Baker and L. DeMarco, *Preperiodic points and unlikely intersections*, Duke Math. J. **159** (2011), no. 1, 1–29.
4. M. Baker and R. Rumely, *Potential theory and dynamics on the Berkovich projective line*, Math. Surveys Monogr., vol. 159, Amer. Math. Soc., Providence, RI, 2010.
5. A. Baragar, *Rational points on K3 surfaces in* $\mathbf{P}^1 \times \mathbf{P}^1 \times \mathbf{P}^1$, Math. Ann. **305** (1996), no. 3, 541–558.
6. ———, *Rational curves on K3 surfaces in* $\mathbf{P}^1 \times \mathbf{P}^1 \times \mathbf{P}^1$, Proc. Amer. Math. Soc. **126** (1998), no. 3, 637–644.
7. ———, *Canonical vector heights on algebraic K3 surfaces with Picard number two*, Canad. Math. Bull. **46** (2003), no. 4, 495–508.
8. ———, *Orbits of curves on certain K3 surfaces*, Compositio Math. **137** (2003), no. 2, 115–134.
9. ———, *Canonical vector heights on K3 surfaces with Picard number three—an argument for nonexistence*, Math. Comp. **73** (2004), no. 248, 2019–2025 (electronic).
10. A. Baragar and R. van Luijk, *K3 surfaces with Picard number three and canonical vector heights*, Math. Comp. **76** (2007), no. 259, 1493–1498 (electronic).
11. P.-G. Becker, *Transcendence of the values of functions satisfying generalized Mahler type functional equations*, J. Reine Angew. Math. **440** (1993), 111–128.
12. P.-G. Becker and W. Bergweiler, *Transcendency of local conjugacies in complex dynamics and transcendency of their values*, Manuscripta Math. **81** (1993), no. 3-4, 329–337.
13. R. L. Benedetto, *Heights and preperiodic points of polynomials over function fields*, Int. Math. Res. Not. **62** (2005), 3855–3866.
14. D. Bertrand, *Problèmes de transcendance liés aux hauteurs sur les courbes elliptiques*, Mathématiques, CTHS: Bull. Sec. Sci., vol. 3, Bib. Nat., Paris, 1981, pp. 55–63.
15. A. Bhatnagar, *Points of canonical height zero on projective varieties* (2010).
16. A. Bhatnagar and L. Szpiro, *Canonical height zero over function fields*, in preparation.
17. E. Bombieri and W. Gubler, *Heights in Diophantine geometry*, New Math. Monogr., vol. 4, Cambridge Univ. Press, Cambridge, 2006.
18. E. Bombieri, D. Masser, and U. Zannier, *Intersecting a curve with algebraic subgroups of multiplicative groups*, Internat. Math. Res. Notices **20** (1999), 1119–1140.
19. ———, *Anomalous subvarieties—structure theorems and applications*, Int. Math. Res. Not. IMRN **19** (2007), Art. ID rnm057, 33 pp.
20. ———, *On unlikely intersections of complex varieties with tori*, Acta Arith. **133** (2008), no. 4, 309–323.
21. T. Bousch, *Sur quelques problèmes de dynamique holomorphe*, Ph.D. Thesis, Université Paris-Sud 11, 1992.
22. J.-Y. Briend and J. Duval, *Exposants de Liapounoff et distribution des points périodiques d'un endomorphisme de* $\mathbf{C}\mathrm{P}^k$, Acta Math. **182** (1999), no. 2, 143–157.
23. X. Buff and A. Epstein, *Bifurcation measure and postcritically finite rational maps*, Complex Dynamics (D. Schleicher, ed.), A K Peters, Wellesley, MA, 2009, pp. 491–512.
24. X. Buff and L. Tan, *The quadratic dynatomic curves are smooth and irreducible*, preprint.
25. L. Busé, *Elimination theory in codimension one and applications*, Technical Report 5918, INRIA, 2006.

26. G. S. Call and J. H. Silverman, *Canonical heights on varieties with morphisms*, Compositio Math. **89** (1993), no. 2, 163–205.
27. A. Chambert-Loir and A. Thuillier, *Mesures de Mahler et équidistribution logarithmique*, Ann. Inst. Fourier (Grenoble) **59** (2009), no. 3, 977–1014.
28. Z. Chatzidakis and E. Hrushovski, *Difference fields and descent in algebraic dynamics*. I, J. Inst. Math. Jussieu **7** (2008), no. 4, 653–686; II, 687–704.
29. D. A. Cox, J. Little, and D. O'Shea, *Using algebraic geometry*, 2nd ed., Grad. Texts in Math., vol. 185, Springer, New York, 2005.
30. L. DeMarco, *Dynamics of rational maps: Lyapunov exponents, bifurcations, and capacity*, Math. Ann. **326** (2003), no. 1, 43–73.
31. L. Denis, *Points périodiques des automorphismes affines*, J. Reine Angew. Math. **467** (1995), 157–167.
32. I. Dolgachev, *Lectures on invariant theory*, London Math. Soc. Lecture Note Ser., vol. 296, Cambridge Univ. Press, Cambridge, 2003.
33. A. Douady and J. H. Hubbard, *Exploring the Mandelbrot set. The Orsay notes*, available at `www.math.cornell.edu/~hubbard/OrsayEnglish.pdf`.
34. ______, *Itération des polynômes quadratiques complexes*, C. R. Acad. Sci. Paris Sér. I Math. **294** (1982), no. 3, 123–126.
35. ______, *Étude dynamique des polynômes complexes. Partie* II, with the collaboration of P. Lavaurs, Tan Lei, and P. Sentenac, Publ. Math. Orsay, vol. 85, Univ. Paris-Sud, Dép. de Mathématiques, Orsay, 1985.
36. ______, *A proof of Thurston's topological characterization of rational functions*, Acta Math. **171** (1993), no. 2, 263–297.
37. A. Epstein, *Integrality and rigidity for postcritically finite polynomials*, available at `arXiv:1010.2780`.
38. G. Everest and T. Ward, *Heights of polynomials and entropy in algebraic dynamics*, Universitext, Springer, London, 1999.
39. N. Fakhruddin, *Boundedness results for periodic points on algebraic varieties*, Proc. Indian Acad. Sci. Math. Sci. **111** (2001), no. 2, 173–178.
40. ______, *Questions on self maps of algebraic varieties*, J. Ramanujan Math. Soc. **18** (2003), no. 2, 109–122.
41. G. Faltings, *Finiteness theorems for abelian varieties over number fields*, Arithmetic Geometry (Storrs, CT, 1984), Springer, New York, 1986, pp. 9–27. Translated from the German original [Invent. Math. **73** (1983), no. 3, 349–366; Erratum, **75** (1984), no. 2, 381] by E. Shipz.
42. E. V. Flynn, B. Poonen, and E. F. Schaefer, *Cycles of quadratic polynomials and rational points on a genus-2 curve*, Duke Math. J. **90** (1997), no. 3, 435–463.
43. D. Ghioca, L.-C. Hsia, and T. J. Tucker, *Preperiodic points for families of polynomials*, available at `arXiv:1102.2769`.
44. D. Ghioca and Michael. E. Zieve, *Lattès maps in arbitrary characteristic*, in preparation.
45. W. Green, *Heights in families of abelian varieties*, Duke Math. J. **58** (1989), no. 3, 617–632.
46. P. Habegger, *Intersecting subvarieties of abelian varieties with algebraic subgroups of complementary dimension*, Invent. Math. **176** (2009), no. 2, 405–447.
47. ______, *On the bounded height conjecture*, Int. Math. Res. Not. IMRN **5** (2009), 860–886.
48. M. Hindry and J. H. Silverman, *Diophantine geometry: An introduction*, Grad. Texts in Math., vol. 201, Springer, New York, 2000.
49. B. Hutz, *Good reduction of periodic points on projective varieties*, Illinois J. Math. **53** (2009), no. 4, 1109–1126.
50. ______, *Dynatomic cycles for morphisms of projective varieties*, New York J. Math. **16** (2010), 125–159.
51. ______, *Effectivity of dynatomic cycles for morphisms of projective varieties using deformation theory*, Proc. Amer. Math. Soc., to appear, available at `arXiv:1011.5155`.
52. D. Huybrechts, *Compact hyperkähler manifolds*, Calabi–Yau Manifolds and Related Geometries (Nordfjordeid, 2001), Universitext, Springer, Berlin, 2003, pp. 161–225.
53. P. Ingram, *A finiteness result for post-critically finite polynomials*, Int. Math. Res. Not. IMRN, to appear, available at `arXiv:1010.3393`.
54. ______, *Variation of the canonical height for a family of polynomials*, available at `arXiv:1003.4225`.
55. ______, *Canonical heights for Hénon maps*, preprint.

56. J.-P. Jouanolou, *Le formalisme du résultant*, Adv. Math. **90** (1991), no. 2, 117–263.
57. S. Kawaguchi, *Canonical height functions for affine plane automorphisms*, Math. Ann. **335** (2006), no. 2, 285–310.
58. ———, *Local and global canonical height functions for affine space regular automorphisms*, available at `arXiv:0909.3573`.
59. S. Kawaguchi and J. H. Silverman, *Canonical heights and the arithmetic complexity of morphisms on projective space*, Pure Appl. Math. Q. **5** (2009), no. 4, 1201–1217. Special issue in honor of John Tate.
60. ———, *Nonarchimedean Green functions and dynamics on projective space*, Math. Z. **262** (2009), no. 1, 173–197.
61. A. Khrennikov, *Non-Archimedean analysis: quantum paradoxes, dynamical systems and biological models*, Math. Appl., vol. 427, Kluwer, Dordrecht, 1997.
62. S. Lang, *Fundamentals of Diophantine geometry*, Springer, New York, 1983.
63. E. Lau and D. Schleicher, *Internal addresses in the Mandelbrot set and irreducibility of polynomials*, Technical Report ims94-19, Stony Brook Institute for Mathematical Sciences, 1994.
64. C. Lee, *An upper bound for the height for regular affine automorphisms of* $\mathbb{A}^n$, available at `arXiv:0909.3107`.
65. ———, *Height estimates for rational maps*, ProQuest LLC, Ann Arbor, MI, 2010. Ph.D. Thesis, Brown University.
66. A. Levy, *The space of morphisms on projective space*, Acta Arith. **146** (2011), no. 1, 13–31.
67. R. Mañé, P. Sad, and D. Sullivan, *On the dynamics of rational maps*, Ann. Sci. École Norm. Sup. (4) **16** (1983), no. 2, 193–217.
68. M. Manes, *$\mathbb{Q}$-rational cycles for degree-2 rational maps having an automorphism*, Proc. Lond. Math. Soc. (3) **96** (2008), no. 3, 669–696.
69. S. Marcello, *Sur la dynamique arithmétique des automorphismes affines*, Ph.D. thesis, Université Paris 7, 2000.
70. ———, *Sur les propriétés arithmétiques des itérés d'automorphismes réguliers*, C. R. Acad. Sci. Paris Sér. I Math. **331** (2000), no. 1, 11–16.
71. D. Masser and U. Zannier, *Torsion anomalous points and families of elliptic curves*, C. R. Math. Acad. Sci. Paris **346** (2008), no. 9-10, 491–494.
72. B. Mazur, *Modular curves and the Eisenstein ideal*, Inst. Hautes Études Sci. Publ. Math. **47** (1977), 33–186 (1978).
73. C. T. McMullen, *Families of rational maps and iterative root-finding algorithms*, Ann. of Math. (2) **125** (1987), no. 3, 467–493.
74. ———, *Complex dynamics and renormalization*, Ann. of Math. Stud., vol. 135, Princeton Univ. Press, Princeton, NJ, 1994.
75. ———, *Dynamics on $K3$ surfaces: Salem numbers and Siegel disks*, J. Reine Angew. Math. **545** (2002), 201–233.
76. C. T. McMullen and D. P. Sullivan, *Quasiconformal homeomorphisms and dynamics. III: The Teichmüller space of a holomorphic dynamical system*, Adv. Math. **135** (1998), no. 2, 351–395.
77. L. Merel, *Bornes pour la torsion des courbes elliptiques sur les corps de nombres*, Invent. Math. **124** (1996), no. 1-3, 437–449.
78. J. Milnor, *Geometry and dynamics of quadratic rational maps*, with an appendix by J. Milnor and L. Tan, Experiment. Math. **2** (1993), no. 1, 37–83.
79. ———, *On rational maps with two critical points*, Experiment. Math. **9** (2000), no. 4, 481–522.
80. ———, *On Lattès maps*, Dynamics on the Riemann Sphere (Holbæk, 2003) (P. G. Hjorth and C. L. Petersen, eds.), Eur. Math. Soc., Zürich, 2006, pp. 9–43.
81. A. Mimar, *On the preperiodic points of an endomorphism of $\mathbb{P}^1 \times \mathbb{P}^1$ which lie on a curve*, Ph.D. Thesis, Columbia University, 1997.
82. P. Morton, *Arithmetic properties of periodic points of quadratic maps*, Acta Arith. **62** (1992), no. 4, 343–372.
83. ———, *On certain algebraic curves related to polynomial maps*, Compositio Math. **103** (1996), no. 3, 319–350.
84. P. Morton and J. H. Silverman, *Rational periodic points of rational functions*, Internat. Math. Res. Notices **2** (1994), 97–110.

85. D. Mumford, *On the equations defining abelian varieties.* I, Invent. Math. **1** (1966), 287–354.
86. ______, *On the equations defining abelian varieties.* II, Invent. Math. **3** (1967), 75–135.
87. ______, *On the equations defining abelian varieties.* III, Invent. Math. **3** (1967), 215–244.
88. D. Mumford, J. Fogarty, and F. Kirwan, *Geometric invariant theory*, 3rd ed., Ergeb. Math. Grenzgeb. (2), vol. 34, Springer, Berlin, 1994.
89. D. Mumford and K. Suominen, *Introduction to the theory of moduli*, Algebraic Geometry, Oslo, 1970 (Oslo, 1970) (F. Oort, ed.), Wolters-Noordhoff, Groningen, 1972, pp. 171–222.
90. W. Narkiewicz, *Polynomial mappings*, Lecture Notes in Math., vol. 1600, Springer, Berlin, 1995.
91. D G. Northcott, *Periodic points on an algebraic variety*, Ann. of Math. (2) **51** (1950), 167–177.
92. Y. Okuyama, *Repelling periodic points and logarithmic equidistribution in non-archimedean dynamics*, available at `arXiv:1106.3363`.
93. F. Oort, *Canonical liftings and dense sets of CM-points*, Arithmetic Geometry (Cortona, 1994) (F. Catanese, ed.), Sympos. Math., vol. 37, Cambridge Univ. Press, Cambridge, 1997, pp. 228–234.
94. F. Pakovich, *Conservative polynomials and yet another action of* $\mathrm{Gal}(\overline{\mathbb{Q}}/\mathbb{Q})$ *on plane trees*, J. Théor. Nombres Bordeaux **20** (2008), no. 1, 205–218.
95. C. Petsche, L. Szpiro, and M. Tepper, *Isotriviality is equivalent to potential good reduction for endomorphisms of* $\mathbb{P}^N$ *over function fields*, J. Algebra **322** (2009), no. 9, 3345–3365.
96. C. Petsche, L. Szpiro, and T. J. Tucker, *A dynamical pairing between two rational maps*, Trans. Amer. Math. Soc., to appear, available at `arXiv:0911.1875`.
97. J. Pila, *O-minimality and the André–Oort conjecture for* $\mathbb{C}^n$, Ann. of Math. (2) **173** (2011), no. 3, 1779–1840.
98. R. Pink, *A common generalization of the conjectures of André–Oort, Manin–Mumford, and Mordell–Lang* (2005), unpublished.
99. K. Schmidt, *Dynamical systems of algebraic origin*, Progr. Math., vol. 128, Birkhäuser, Basel, 1995.
100. J.-P. Serre, *Local fields*, Grad. Texts in Math., vol. 67, Springer, New York, 1979. Translated from the French by M. J. Greenberg.
101. J.-P. Serre and J. Tate, *Good reduction of abelian varieties*, Ann. of Math. (2) **88** (1968), 492–517.
102. N. Sibony, *Dynamique des applications rationnelles de* $\mathbf{P}^k$, Dynamique et géométrie complexes (Lyon, 1997), Panor. Synthèses, vol. 8, Soc. Math. France, Paris, 1999, pp. ix–x, xi–xii, 97–185.
103. J. H. Silverman, *Heights and the specialization map for families of abelian varieties*, J. Reine Angew. Math. **342** (1983), 197–211.
104. ______, *Computing heights on elliptic curves*, Math. Comp. **51** (1988), no. 183, 339–358.
105. ______, *Rational points on* $K3$ *surfaces: a new canonical height*, Invent. Math. **105** (1991), no. 2, 347–373.
106. ______, *Geometric and arithmetic properties of the Hénon map*, Math. Z. **215** (1994), no. 2, 237–250.
107. ______, *The field of definition for dynamical systems on* $\mathbf{P}^1$, Compositio Math. **98** (1995), no. 3, 269–304.
108. ______, *Computing canonical heights with little (or no) factorization*, Math. Comp. **66** (1997), no. 218, 787–805.
109. ______, *The space of rational maps on* $\mathbf{P}^1$, Duke Math. J. **94** (1998), no. 1, 41–77.
110. ______, *The arithmetic of dynamical systems*, Grad. Texts in Math., vol. 241, Springer, New York, 2007.
111. ______, *The arithmetic of elliptic curves*, 2nd ed., Grad. Texts in Math., vol. 106, Springer, Dordrecht, 2009.
112. ______, *Height estimates for equidimensional dominant rational maps*, J. Ramanujan Math. Soc. **26** (2011), no. 2, 145–163.
113. ______, *An algebraic approach to certain cases of Thurston rigidity*, Proc. Amer. Math. Soc., to appear, available at `arXiv:1010.4562`.
114. D. Speyer, *Rational maps with all critical points fixed*, MathOverflow, `mathoverflow.net/questions/4983`.

115. M. Stoll, *Rational* 6*-cycles under iteration of quadratic polynomials*, LMS J. Comput. Math. **11** (2008), 367–380.
116. L. Szpiro and T. J. Tucker, *A Shafarevich–Faltings theorem for rational functions*, Pure Appl. Math. Q. **4** (2008), no. 3, 715–728.
117. J. Tate, *Variation of the canonical height of a point depending on a parameter*, Amer. J. Math. **105** (1983), no. 1, 287–294.
118. D. Tischler, *Critical points and values of complex polynomials*, J. Complexity **5** (1989), no. 4, 438–456.
119. C. Voisin, *Géométrie des espaces de modules de courbes et de surfaces* $K3$ (*d'après Gritsenko-Hulek-Sankaran, Farkas-Popa, Mukai, Verra, et al.*), Astérisque **317** (2008), x, 467–490. Séminaire Bourbaki. Vol. 2006/2007, Exp. No. 981.
120. M. Waldschmidt, *Algebraic dynamics and transcendental numbers*, Noise, Oscillators and Algebraic Randomness (Chapelle des Bois, 1999) (M. Planat, ed.), Lecture Notes in Phys., vol. 550, Springer, Berlin, 2000, pp. 372–378.
121. L. Wang, *Rational points and canonical heights on* $K3$*-surfaces in* $\mathbf{P}^1 \times \mathbf{P}^1 \times \mathbf{P}^1$, Recent developments in the inverse Galois problem (Seattle, WA, 1993) (M. D. Fried, S. S. Abhyankar, W. Feit, Y. Ihara, and H. Völklein, eds.), Contemp. Math., vol. 186, Amer. Math. Soc., Providence, RI, 1995, pp. 273–289.
122. S. Zhang, *Small points and adelic metrics*, J. Algebraic Geom. **4** (1995), no. 2, 281–300.
123. M. E. Zieve, *Cycles of polynomial mappings*, ProQuest LLC, Ann Arbor, MI, 1996. Ph.D. Thesis, University of California, Berkeley.
124. B. Zilber, *Exponential sums equations and the Schanuel conjecture*, J. London Math. Soc. (2) **65** (2002), no. 1, 27–44.
125. K. Zsigmondy, *Zur Theorie der Potenzreste*, Monatsh. Math. Phys. **3** (1892), no. 1, 265–284.

Additional References

A reasonably up-to-date list of references for arithmetic dynamics is maintained at

`http://math.brown.edu/~jhs/ADSHome.html`

Index

Titles in This Series

30 **Joseph H. Silverman,** Moduli spaces and arithmetic dynamics, 2012

29 **Marcelo Aguiar and Swapneel Mahajan,** Monoidal functors, species and Hopf algebras, 2010

28 **Saugata Ghosh,** Skew-orthogonal polynomials and random matrix theory, 2009

27 **Jean Berstel, Aaron Lauve, Christophe Reutenauer, and Franco V. Saliola,** Combinatorics on words: Christoffel words and repetitions in words, 2009

26 **Victor Guillemin and Reyer Sjamaar,** Convexity properties of Hamiltonian group actions, 2005

25 **Andrew J. Majda, Rafail V. Abramov, and Marcus J. Grote,** Information theory and stochastics for multiscale nonlinear systems, 2005

24 **Dana Schlomiuk, Andreĭ A. Bolibrukh, Sergei Yakovenko, Vadim Kaloshin, and Alexandru Buium,** On finiteness in differential equations and Diophantine geometry, 2005

23 **J. J. M. M. Rutten, Marta Kwiatkowska, Gethin Norman, and David Parker,** Mathematical techniques for analyzing concurrent and probabilistic systems, 2004

22 **Montserrat Alsina and Pilar Bayer,** Quaternion orders, quadratic forms, and Shimura curves, 2004

21 **Andrei Tyurin,** Quantization, classical and quantum field theory and theta functions, 2003

20 **Joel Feldman, Horst Knörrer, and Eugene Trubowitz,** Riemann surfaces of infinite genus, 2003

19 **L. Lafforgue,** Chirurgie des grassmanniennes, 2003

18 **G. Lusztig,** Hecke algebras with unequal parameters, 2003

17 **Michael Barr,** Acyclic models, 2002

16 **Joel Feldman, Horst Knörrer, and Eugene Trubowitz,** Fermionic functional integrals and the renormalization group, 2002

15 **José I. Burgos Gil,** The regulators of Beilinson and Borel, 2002

14 **Eyal Z. Goren,** Lectures on Hilbert modular varieties and modular forms, 2002

13 **Michael Baake and Robert V. Moody, Editors,** Directions in mathematical quasicrystals, 2000

12 **Masayoshi Miyanishi,** Open algebraic surfaces, 2001

11 **Spencer J. Bloch,** Higher regulators, algebraic K-theory, and zeta functions of elliptic curves, 2000

10 **James D. Lewis,** A survey of the Hodge conjecture, Second Edition, 1999

9 **Yves Meyer,** Wavelets, vibrations and scaling, 1998

8 **Ioannis Karatzas,** Lectures on the mathematics of finance, 1996

7 **John Milton,** Dynamics of small neural populations, 1996

6 **Eugene B. Dynkin,** An introduction to branching measure-valued processes, 1994

5 **Andrew Bruckner,** Differentiation of real functions, 1994

4 **David Ruelle,** Dynamical zeta functions for piecewise monotone maps of the interval, 1994

3 **V. Kumar Murty,** Introduction to Abelian varieties, 1993

2 **M. Ya. Antimirov, A. A. Kolyshkin, and Rémi Vaillancourt,** Applied integral transforms, 1993

1 **D. V. Voiculescu, K. J. Dykema, and A. Nica,** Free random variables, 1992